A Field Guide to Edible Wild Plants

of Eastern and Central North America

THE PETERSON FIELD GUIDE SERIES®
Edited by Roger Tory Peterson

Advanced Birding—*Kaufman*
Birds of Britain and Europe—*Peterson, Mountfort, Hollom*
Birds of Eastern and Central North America—*R.T. Peterson*
Birds of Texas and Adjacent States—*R.T. Peterson*
Birds of the West Indies—*Bond*
Eastern Birds' Nests—*Harrison*
Hawks—*Clark and Wheeler*
Hummingbirds—Williamson
Mexican Birds—*R.T. Peterson and Chalif*
Warblers—Dunn and Garrett
Western Birds—*R.T. Peterson*
Western Birds' Nests—*Harrison*
Backyard Bird Song—*Walton and Lawson*
Eastern Bird Songs —*Cornell Laboratory of Ornithology*
Eastern Birding by Ear—*Walton and Lawson*
More Birding by Ear: Eastern and Central—*Walton and Lawson*
Western Bird Songs—*Cornell Laboratory of Ornithology*
Western Birding by Ear—*Walton and Lawson*
Pacific Coast Fishes—*Eschmeyer, Herald, and Hammann*
Atlantic Coast Fishes—*Robins, Ray, and Douglass*
Freshwater Fishes (N. America north of Mexico)—*Page and Burr*
Insects (America north of Mexico)—*Borror and White*
Beetles—*White*
Eastern Butterflies—*Opler and Malikul*
Western Butterflies—*Opler and Wright*
Mammals—*Burt and Grossenheider*
Animal Tracks—*Murie*
Eastern Forests—*Kricher and Morrison*
California and Pacific Northwest Forests—Kricher and Morrison
Rocky Mountain and Southwest Forests—Kricher and Morrison
Venomous Animals and Poisonous Plants—Foster and Caras
Edible Wild Plants (e. and cen. N. America)—*L. Peterson*
Eastern Medicinal Plants and Herbs—*Foster and Duke*
Eastern Trees—*Petrides*
Ferns (ne. and cen. N. America)—*Cobb*
Mushrooms—*McKnight and McKnight*
Pacific States Wildflowers—*Niehaus and Ripper*
Western Medicinal Plants and Herbs—*Foster and Hobbs*
Rocky Mt. Wildflowers—*Craighead, Craighead, and Davis*
Trees and Shrubs—*Petrides*
Western Trees—Petrides
Wildflowers (ne. and n.-cen. N. America)—*R.T. Peterson and McKenney*
Southwest and Texas Wildflowers—*Niehaus, Ripper, and Savage*
Geology (e. N. America)—*Roberts*
Rocks and Minerals—*Pough*
Stars and Planets—*Pasachoff*
Atmosphere—*Schaefer and Day*
Eastern Reptiles and Amphibians—*Conant and Collins*
Western Reptiles and Amphibians—*Stebbins*
Shells of the Atlantic and Gulf Coasts, W. Indies—*Morris*
Pacific Coast Shells (including Hawaii)—*Morris*
Atlantic Seashore—*Gosner*
Coral Reefs (Caribbean and Florida)—*Kaplan*
Southeastern and Caribbean Seashores—*Kaplan*

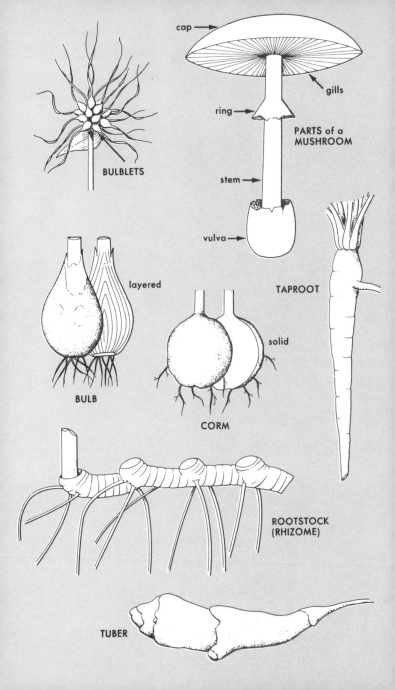

BULBLETS

PARTS of a MUSHROOM

cap

gills

ring

stem

vulva

TAPROOT

BULB

layered

CORM

solid

ROOTSTOCK
(RHIZOME)

TUBER

THE PETERSON FIELD GUIDE SERIES®

A Field Guide to
Edible
Wild Plants

Eastern and central North America

Lee Allen Peterson

Line Drawings by
Lee Allen Peterson
and
Roger Tory Peterson

Photographs by
Lee Allen Peterson

*Sponsored by the National Audubon Society
and the National Wildlife Federation*

HOUGHTON MIFFLIN COMPANY
Boston   New York

Copyright © 1977 by Lee Peterson

PETERSON FIELD GUIDES and PETERSON FIELD GUIDE SERIES
are registered trademarks of Houghton Mifflin Company.

Library of Congress Cataloging in Publication Data

Peterson, Lee.
A field guide to edible wild plants of
Eastern and Central North America.

(The Peterson field guide series; no. 23)
Includes index.
1. Wild plants, Edible—United States—Identification.
2. Wild plants, Edible—Canada—Identification.
I. Title.
QK98.5.U6P47   581.6′32′0973   77-27323
ISBN 0-395-20445-3 (hardbound)
ISBN 0-395-92622-X

Printed in the United States of America

VB   34 33 32 31 30 29 28 27 26 25

Visit our Web site: www.houghtonmifflinbooks.com.

Editor's Note

EARLY MAN knew a great deal about wild plants, the ones he could eat and those he could not. Actually, like the other anthropoids, he was probably a vegetarian—until he learned to hunt.

Much of this native knowledge of wild plants was lost when men learned to cultivate the land and to harvest crops. But some of the lore still persists among country people to this day. When I was a boy in western New York State, the Italian women from down the hill came to glean the dandelions from our lawn, picking the leaves to boil as greens, and snipping the flowerheads to brew into dandelion wine. We Swedish-Americans found dandelion greens a trifle too bitter for our taste and much preferred the leaves of marshmarigolds, or "cowslips." In the early spring, we made frequent excursions to the swampy spots north of town to gather basketfuls of the fresh new leaves (which, by the way, are poisonous if eaten without first being cooked). To make wine, some of our neighbors picked elderberries, but my mother preferred to can them, filling our cellar with quart mason jars of these fruits which eventually went into delicious pies.

In this era of the supermarket, few people still take advantage of the great store of edible wild plants available to them. Perhaps it is too much bother. But with escalating prices of market products, more and more people are rediscovering the wild foods that are free for the taking. Or perhaps this is a spinoff of the recent breakthrough toward environmental awareness. Campers, hikers, and especially those young people in blue jeans who were on limited budgets revived the culinary art of the roadside, trailside, and wilderness. The late Euell Gibbons was their high priest.

My son, Lee, was already deep into gastronomic botany when he graduated from Johns Hopkins University. Although he found a number of publications that extolled the joys of eating on the wild side, he discovered that most of them dealt primarily with recipes; they did not go much into field identification. To partially fill this need, he decided to produce a pamphlet on the edible wild plants of Connecticut and distribute it through local health food stores. I urged him to expand it to cover the eastern half of the United States and Canada and have it published in book form. This Field Guide is the end result of his labors. Some of the line drawings are mine, adapted from *A Field Guide to Wildflowers of Northeastern and North-central North America*. Many of the others, particularly those of southern plants as well as various tubers, roots, fruits, and the like that were not illustrated in the flower guide, were carefully drawn by Lee. Almost all of the photographs in the color section are his and are the product of his extensive travel throughout the East.

Obviously, "Nature's free harvest" alone cannot adequately feed

our burgeoning populations. Even around our small town of Old
Lyme, Connecticut, the edible plants of the nearby woods,
marshes, and fields would soon be depleted if everyone depended
on them for survival. But as more and more people take to the
trails, or go canoeing into the back country, a knowledge of what
can be eaten and what cannot should be of great value. The
camper can travel more lightly; he can vary his diet with items
that have a different taste and that frequently have a higher
vitamin content than commercially grown vegetables.

On your next camping trip, slip this Field Guide into your
backpack and study it at your leisure. If you become lost or
separated from your store of provisions, you need not starve. Even
a little knowledge about edible wild plants will get you out alive
and healthy.

ROGER TORY PETERSON

Preface

In 1968 I spent the summer working at Camp Chewonki, a boy's camp on the Maine coast that stresses natural history and woodcraft skills. Up until that time my experience with edible wild plants had been limited to picking an occasional raspberry or blueberry. When I arrived at camp, a friend introduced me to Euell Gibbons' *Stalking the Wild Asparagus*. That book, together with my father's newly released *Field Guide to Wildflowers of Northeastern and North-central North America,* opened up a whole new level of discovery for me.

I get much the same pleasure from foraging for wild foods as I do from fly-fishing for trout. Fishing provides a relaxing break when work becomes onerous and is an excuse to explore the out-of-doors. Small fast-flowing streams are my favorite haunts; I enjoy reading the currents and eddies and being able to predict where the fish will be. The unsullied surroundings and rhythmic casting motions make the experience extraordinarily peaceful. Yet, at the same time there is always that undercurrent of excitement—waiting for the fish to strike. Gathering edible wild plants evokes many of the same responses. Here, too, I become immersed in my surroundings, acutely aware of the plants and the processes that affect them. Every element conveys a message that is both familiar and revealing. Together, they enable me to anticipate what edible plants will be present, and where. Of course, the enjoyment does not end with the discovery—or in fishing with the catch. Beyond lies the satisfaction of preparing the food and eating it.

There are a variety of other reasons for becoming involved with wild foods. They are a legitimate way of economizing on food costs; many of the more abundant and nutritious species probably grow within a short walking distance of your home. They are also an excellent source of vitamins and minerals. This is not to say that plants from the wild are necessarily richer in vital nutrients than those grown in a garden, but they are almost always healthier than store-bought vegetables. Commercially grown vegetables are bred for size and appearance, and for this reason their flavor and nutritive value often suffer. Furthermore, by the time they have been processed, shipped, and displayed for a few days in a supermarket, many of their vitamins and minerals have been lost. Wild vegetables, on the other hand, are usually eaten shortly after they are gathered and, when properly prepared, retain most of their original flavor and food value. Backpackers can use them to extend supplies or lighten their load. And, in rare instances, a knowledge of edible wild plants may even be essential for survival; it is still possible to be lost in eastern North America, or otherwise unable to obtain food from more conventional sources.

You can use wild foods to varying degrees, ranging all the way from living on them entirely to occasionally supplementing your regular meals with them. I generally treat foraging for wild foods as a form of outdoor therapy and experiment with new species whenever it strikes my fancy. However, regardless of how frequently you use them, the ability to feed yourself in virtually any situation gives you an extraordinary sense of freedom and self-reliance.

Some of the benefits to be derived from foraging for edible plants are obvious. Others are less readily apparent. As you become more aware of your surroundings, you make many small, interesting discoveries. You get a feeling for the cycles of nature—for instance, the flow of energy in plants. The edible parts to be collected at any particular time are always those that have the highest concentrations of food energy—food energy for the plant as well as for you. In the spring and summer, this means that the leafy shoots are where the energy is concentrated; in the fall, the seeds and fruit; and in the winter months, the roots. Most important, however, collecting edible plants gives you a greater insight into the processes that are taking place around you. When you are looking for a particular plant, it becomes critical to know where it will occur, and why. Because of this, the interrelationships between the inorganic and organic worlds become much more clearly defined. A far subtler influence results from the extraordinary range of tastes and textures inherent in wild plants. A diet restricted to a set number of foods is just as harmful as any other type of physical or spiritual confinement and experimenting with edible wild plants can do much to dispel food biases.

A great many people have contributed, directly or indirectly, to the production of this book. Dr. David S. Barrington, Curator of the Pringle Herbarium, and Dr. Hubert W. Vogelmann of the Department of Botany at the University of Vermont, made both themselves and their resources available to me; the University provided me with office space on the Burlington campus. Jay M. Arena, M.D., Duke Hospital, Durham, North Carolina, and Dr. James W. Hardin, Department of Botany at North Carolina State University, reviewed the short discussion of poisonous plants in the introduction. George A. Petrides graciously consented to the use of illustrations from *A Field Guide to Trees and Shrubs.*

For complete access to their slide collections I would like to thank Dr. and Mrs. Oliver Eastman, Diana Kappel, Dr. Steven Young, and Cecil B. Hoisington. Patty Rosencrantz designed and drew the food type symbols for Asparagus, Pickle, Potato, and made useful suggestions regarding the others.

Karl and Ursula Schaefer gave me a second home during the difficult final stages in the completion of the manuscript. I would also like to thank the following people for their help, support, and friendship: Ellen Adams, Chris and Barb Barrett, Susan Bratton, Forest Buchanan, Bud and Ruth Carr, John and Nancy Christie,

Kelvin Cole, John Henry Dick, Anne Dinsmore, Dr. Greg Lanne Eddy, Dr. Richard B. Fischer, Mr. and Mrs. H. Lincoln Foster, Steve Garber, Dr. William T. Gillis, Dr. John M. Graham, Dr. Richard H. Goodwin, Dr. and Mrs. Joseph Hickey, Louise Hoehl, Barbara Hoisington, William Hoisington, Jane and Russ Kinne, Bob and Katie Lewin, Dr. William A. Niering, Mr. and Mrs. George Poundstone, Charles L. Ripper, Dr. James Rodman, Shapleigh Smith, Elizabeth Spear, Paul Spitzer, Alfred Tizzie, Jr., Muff Wagner, Alex and Jean Vasiloff; also Doyle McKey, who lent me a copy of Euell Gibbons' *Stalking The Wild Asparagus* and so inadvertently began the progression of events that led to this book. Pidge Eastland and Elaine Giambattista typed the bulk of the manuscript; additional typing was done by Isabel O'Brien, Barbara Peterson, and Susan Spitzer. I am extremely grateful to all of them.

My thanks also to the people at Houghton Mifflin Company who have been instrumental in bringing this book to completion. Among them are Morton H. Baker, Katharine R. Bernard, Richard McAdoo, Austin Olney, Carol Goldenberg, Stephen Pekich, James Thompson, and Richard Tonachel. Special thanks go to Virginia Ehrlich, who worked with me in establishing the overall format and style of the Guide, and also did the preliminary copyediting; Lisa Gray Fisher did the final editing and saw the manuscript through the subsequent stages of production.

Finally and perhaps most important, I would like to thank my parents, who gave me their encouragement, support, and guidance.

Contents

Illustrations

THE GEOGRAPHICAL AREA
COVERED BY THIS FIELD GUIDE

Includes the eastern and central states and the adjacent provinces of Canada—roughly the area east of the 97th meridian. The southern half of Florida has been excluded as its flora is more typical of the subtropics.

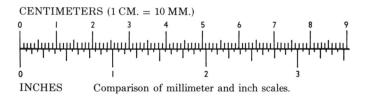

CENTIMETERS (1 CM. = 10 MM.)

INCHES   Comparison of millimeter and inch scales.

How to Use This Book

GATHERING edible wild plants requires a variety of skills beyond simple species recognition. Not only must one know how to identify a plant, but also when and how to find it and what to do with it once it is found.

When you think you have discovered an edible species, run through its identification several times, paying close attention to those features and characteristics that distinguish it from the surrounding plants. Follow it through the seasons until you can recognize it at any stage of its growth. (This is particularly important when dealing with plants that have edible roots or shoots that must be gathered in fall, winter, or early spring when flowers are not available for proper species identification.) Learn its habits so that you can find it again. When and where does it grow? Under what conditions? With what plants is it commonly associated? Learn its uses. What parts are edible and when? How is it prepared? Do not overextend yourself. Concentrate on a dozen or so of the more frequently encountered species in your area at first, and experiment with new ones only when you have become thoroughly familiar with these.

Not all of the edible plants discussed in this guide are noted for their flavor. Some must be suffered through for nutrition's sake during emergencies. Others will not appeal to your particular taste (I can just barely tolerate Brussels sprouts). Make a mental note about these in case of need, but set them aside in favor of more palatable species.

General Organization

This *Field Guide* is divided into three principal sections, each serving a different function:

(1) The *Visual and Descriptive Text* (pp. 17–239) is the largest and most important part of the book and contains all the information necessary for the proper identification, location, and use of the species covered. Each species is illustrated on a right-hand page and has a complete text description facing it on the left. The overall construction of the section serves as a visual key to species identification. Species are grouped into three broad categories according to basic similarities: (1) flowering plants (mainly herbaceous); (2) woody plants (trees, shrubs, vines); and (3) miscellaneous plants (grasslike plants, ferns, seaweeds, lichens, mushrooms). *Flowering plants* are organized according to flower color (indicated at the top right-hand corner of each plate) and subdivided into plates according to the number of petals, arrangement of

1

flowers, shape and position of leaves, and so forth. *Woody plants* are organized according to their leaf-types (needlelike, daggerlike blades, opposite compound, opposite simple, alternate compound, alternate simple) and subdivided into plates according to the size and habit of the mature plants, shapes of leaves, type of fruit, and so on. *Miscellaneous plants* make up the smallest category; I have selected only the few most palatable and readily available varieties.

(2) *Finding Edible Plants* (pp. 241–85) describes the various habitats found in eastern North America in rough successional sequence (open water to dry land, disturbed soil to mature forest) and lists by season the available food plants for each habitat. Abundant, frequently used species are indicated in boldface.

(3) *Food Uses* (pp. 286–312) gives general information on food preparation and lists by season the species that fall within each of the major food-type categories. The food-types are ordered alphabetically. Abundant, frequently used species are indicated in boldface.

Area Covered: This guide includes all the states east of the 97th meridian (except for the southern half of Florida) and the adjacent provinces of Canada from southwestern Ontario to the Gaspé Peninsula (see map, p. xiv). Species coverage is less comprehensive toward the western and southern periphery, where species more typical of the West and subtropics enter our range. Readers seeking additional information on edible plants in the deep South should refer to Morton's *Wild Plants for Survival in South Florida;* and in the West, Kirk's *Wild Edible Plants of the Western United States* (see *Recommended Books,* p. 313).

Species Included: Well over 1000 species of edible plants grow wild in eastern North America. Because space limitations make it impractical to treat all of them in this guide, discussion has been limited to 373 of the more important and better known species. Information on edible species not covered by this guide can be found in Fernald, Kinsey, and Rollin's *Edible Wild Plants of Eastern North America* (see *Recommended Books,* p. 313).

Poisonous Plants: In addition to the discussions and listings on pp. 7–10, 37 species of poisonous plants are treated in detail in the *Visual and Descriptive Text.* For the most part, these are species that might be confused with edible species.

Terms and Symbols: The language in this guide has been kept relatively simple, but a few technical terms were unavoidable. Most of these occur in the descriptive text and refer to specific parts of plants. They are defined in the *Glossary* (pp. 13–14) and/or illustrated on the front or back endpapers.

Symbols are employed in the margins of the text to point out food uses and poisonous species. The definitions for the terms used to denote food types can be found in *Explanation of Symbols* (pp. 15–16). When food types are given as "asparagus," "coffee," and "potato," the terms are used in a general sense to indicate wild species that can be prepared in the same way as these domesticated plants.

Illustrations: The individual drawings are the work of both my father (borrowed from *A Field Guide to Wildflowers* by Roger Tory Peterson and Margaret McKenny and *A Field Guide to Trees and Shrubs* by George A. Petrides) and me; the design of the plates is my own. An open-line style of illustration is used to highlight each species' characteristic shape and structure. Arrows indicate important diagnostic features, which are usually also *italicized* in the accompanying text. A skull and crossbones symbol designates poisonous species.

The color photographs at the center of the book are organized according to habitat and depict characteristic edible species found in each locale. Except for those provided by other people, photographs were taken with a Nikon F1 body, using a 55 mm Micro-Nikkor lens and Kodachrome 25 film. A small Spirolite flash unit was used for flash shots.

Measurements: Because the United States is in the process of converting to the metric system, all measurements are given in duplicate: feet or inches first, followed by metric equivalents rounded to the first decimal. Exact equivalents are: 25.4 millimeters (mm) or 2.54 centimeters (cm) per inch; 30.48 centimeters or 0.3048 meter (m) per foot. However, since most plants display considerable variation in size, I have used a more convenient though slightly less precise set of equivalents: 25 millimeters or 2.5 centimeters per inch; 30 centimeters or 0.3 meter per foot. See p. xiv for inch–centimeter scale.

Recommended Books: This guide is designed to be comprehensive, but you may at some time find yourself in need of supplementary or corroborative information. For this reason, a list of recommended books has been included on p. 313.

Finding Edible Plants

When you set out to find edible wild plants, you have a choice of ways to begin. You can either look at the plants in your vicinity and then try to find them in the text, or you can select a particular species and go in search of it. The former method is slower, more chancy, but could provide some tasty discoveries. With either approach, though, it would be helpful to have a quick checklist of the plants you might expect to find in the area of your search within a given season.

The section entitled *Finding Edible Plants* (pp. 241–85) is designed to serve this purpose. The various habitats are described, and the edible plants found in them are listed by season and ordered by food type. Page numbers refer you to the text descriptions for identification. Concentrate your efforts on the species listed in boldface; these are generally tastier and more readily available.

The boundaries between habitats are sometimes indistinct. This is a natural fuzziness, due to variations in soil and climate, and to the fact that all habitats are in a state of transition. The divisions established for this guide are specific enough to help you locate yourself (and edible plants), but general enough to allow for nature's own flexibility.

Identifying Edible Plants

Species identification in the *Visual and Descriptive Text* is based on visual impressions. Since species are grouped on plates according to obvious similarities, it is usually a simple matter to run through the various divisions and subdivisions until you encounter an illustration that resembles the plant you have in hand. When you think you have a match-up, read the text description of the species, paying close attention to the diagnostic characteristics *italicized* in the text and indicated by arrows on the illustration. For an identification to be positive, the plant should conform to the verbal description in all particulars, including size, habitat, range, and flowering or fruiting dates.

If you are unable to identify a plant or unsure of your identification after double-checking with the text, refer to one of the technical manuals or field guides listed in *Recommended Books* (p. 313). **Do not attempt to use any plant that you cannot positively identify.** Although there are relatively few dangerously poisonous plants growing wild in our area (see list, pp. 8–10), a number of them are somewhat similar in appearance to edible species, and a misidentification could prove unfortunate.

Names: The common and scientific names are given at the beginning of each species entry; the common name(s) is in large boldface type at the top left-hand corner, and the scientific name in italics just below. The term *species* is abbreviated as spp. when it is used in the plural form. Family names are not vital to species identification in this book and have been omitted, lest family relationships be used as a guide to edibility. The common names for wildflowers are from Peterson and McKenny's *A Field Guide to Wildflowers;* the common names for the other plants are those I felt to be in most frequent use and adhere to no one authority. The scientific names for species found in the Northeast are from *Gray's Manual of Botany,* Eighth Edition; the scientific names for species limited to the deep South are from Radford, Ahles, and

Bell's *Manual of the Vascular Flora of the Carolinas* (see *Recommended Books,* p. 313).

Description: Each species' description follows immediately after its scientific name. In cases where several closely related and similar-looking species have the same use, only a general group (genus) description is given at the beginning of the entry, followed by the subentry **Use.** Species are then named and described individually. All text entries should be read with an eye on the illustration.

The measurements that appear at the end of a species description (just before the subentry **Where found**) are plant heights. The figures for flowering plants are the minimum and maximum heights observed for individuals in bloom; those for woody plants are the average heights (and often trunk diameters) of mature specimens.

Where found: Habitats appear first. They rarely need any explanation, but if they should, refer to the appropriate habitat description in *Finding Edible Plants* (pp. 241–85). The limits of a species' range appear next and read from west to east, north to south.

Flowers: Flowering dates represent the period of time in which you can reasonably expect to find a flower in bloom. The dates given cover a species' entire range and may not necessarily represent the actual flowering times for the individuals found in your area. These times show considerable variation, depending on latitude and local changes in altitude, with individuals found at low altitudes or in the South often blooming far earlier than those found at high altitudes or in the North.

Fruit: Fruiting dates are included for species that bear edible fruit, or whose fruit are vital to proper identification. Fruiting times exhibit the same sort of variation as flowering dates (see above).

Using Edible Plants

Virtually all the information needed for the proper preparation and use of each species is contained in its entry in the *Visual and Descriptive Text.*

Parts Used: Edible parts are listed in boldface type at the top right-hand corner of each species entry, on the same line with the common name, and are arranged according to the season in which they become available, starting with spring.

Season of Availability: The season(s) during which each species is edible is given in capital letters at the end of the **Use** portion of the descriptive text. For plants with edible parts that are available at different times of the year, the season of availability for each part is entered separately, along with the name of the part in parentheses.

The divisions of the 5 seasons used in this guide are somewhat arbitrary and vary according to latitude, but they can be generally defined as follows: Spring — the months of March, April, and May; Summer — the months of June, July, and August; Fall — the months of September, October, and November; Fall–Early Spring — September through February; Winter — the months of December, January, and February.

Function of Symbols: The symbols placed in the margin opposite each species' entry in the *Visual and Descriptive Text* indicate its principal uses, given in descending order of importance. Although most of the species discussed in this guide yield a variety of edible products, some of the products are preferable to others. The symbols are designed to serve as a quick reference for readers who seek specific types of food or want to know the best ways to prepare new species. See *Explanation of Symbols*, pp. 15–16, for definitions of the food-types that the symbols represent.

Use and Preparation: Uses and cooking directions are given after the subheading **Use.** If one species has several edible parts requiring different methods of preparation, each set of directions begins with the relevant part in *italics.* The uses listed show the standard ways that a species may be prepared and do not normally include more exotic dishes, such as purées, souffles, and the like. Beers, wines, and vinegars have also been excluded from the list of uses in the text entries, but they are discussed briefly at the end of the section on *Food Uses* (see pp. 311–12).

The cooking methods given in this guide are generally quite simple. They will be enough to give you a start, but you will undoubtedly want to add your own touches as you become more experienced, and more elaborate recipes can be found in the various cookbooks and books on edible plants listed on p. 314. Additional information on general methods of preparation and storage is located under the appropriate food-type headings in the *Food Uses* section (pp. 286–312).

Warning, Caution, Note

These additional boldface subheadings appear in some species accounts in the *Visual and Descriptive Text.* **Warning** is used when a species is potentially poisonous. (Be sure to read the section *Poisonous Plants,* pp. 7–10). **Caution** is used when an edible species might be confused with a poisonous species. **Note** is used for points of special interest, such as edible species that are endangered and in need of protection (see *Conservation,* pp. 11–12).

Poisonous Plants

Do NOT let the fear of being poisoned deter you from experimenting with edible wild plants. Cases of fatal poisoning are extremely rare. Of the thousands of species of plants that grow wild in eastern North America, only a handful can be considered dangerously poisonous (see list for plants found in our area). These should present no problem if you are careful and follow a few simple rules.

1. *Learn to recognize and avoid the common poisonous plants in your area.* Be aware also of those plants that commonly cause dermatitis.

2. *Teach children not to put plants in their mouths; keep all plants away from infants.* Most cases of poisoning involve small children. Parents should learn what poisonous plants grow in or near their homes and warn their children to leave them alone.

3. *Do not use any plant that you cannot positively identify as edible.* If you have even the slightest doubt about the identity of a plant, leave it alone. This is particularly important when dealing with roots, shoots, and berries. Mushrooms (p. 238) and members of the Carrot Family (pp. 38, 40, 42) often defy precise identification and should be approached with extreme caution; mistakes can be fatal.

4. *Do not assume that plants that superficially resemble edible plants are themselves edible.* Unfamiliar members of the Lily Family (6 petal-like flower parts, parallel-veined leaves) and the Pea Family (pealike flowers, pods) may be particularly tempting in this respect. They are just as likely to be poisonous as not.

5. *When collecting an edible plant, make sure not to include parts from nearby poisonous plants.*

6. *Do not collect plants that have recently been sprayed with insecticides, or that grow in contaminated water or along the margins of heavily traveled highways.* Although washing will frequently remove most of the toxic substances from these plants, it is safer and wiser to collect elsewhere.

7. *Be absolutely certain which parts of a plant should be collected and at what season, and the proper way to use them.* Pay close attention to warnings and cautionary notes in the text. Some species are edible when cooked but poisonous when raw, or edible when young but toxic later. Note that certain plants become toxic if eaten in excess.

8. *Sample unfamiliar edible plants sparingly at first.* Body chemistries vary from individual to individual; a plant may be safe for one person to eat, but not for another.

9. *There are no foolproof tests for determining either edible or poisonous plants.* Animals are **not** reliable indicators of edibility.

7

Treatment: Even though most cases of poisoning are much more likely to cause discomfort than to be fatal, quick action is essential. If you suspect that someone has eaten a poisonous plant, *call your doctor immediately.* Be prepared to give him such information as: 1) the name or a good description of the plant; 2) how much of it and which parts were eaten; 3) how long ago it was eaten; 4) age and weight of the patient, and any history of allergies or serious medical problems; 5) symptoms.

If a doctor is unavailable, and the patient is not unconscious or convulsive, have him drink plenty of water and then try to induce vomiting to clear his stomach of any unabsorbed poison. Vomiting can be triggered by gagging the back of the throat with a finger, or by giving an emetic such as syrup of ipecac, mustard water, soapy water, or diluted coffee grounds. Once the patient has vomited, have him continue to drink plenty of liquids and rush him to the nearest hospital or clinic along with a sample of the plant.

If a doctor is unavailable and you cannot get to a hospital, have the patient repeat the vomiting process described above several times until his stomach is thoroughly cleaned out and his vomit is clear. Then, have him drink plenty of liquids and keep him as warm, calm, and comfortable as possible. Should the patient become unconscious or convulsive, do not try to induce vomiting. Keep his mouth and throat free from obstructions (tilting his head back and to the side so that he cannot swallow his tongue or inhale his own vomit) and be prepared to give him mouth-to-mouth resuscitation.

Several books that deal exclusively with poisonous plants are included in the list of *Recommended Books* on p. 313. One or more of these should be kept on hand in case of emergencies.

Plants That Most Commonly Cause Dermatitis

Eyebane
  Euphorbia maculata
Nettles, p. 150
  Urtica spp.
Poison Ivy, p. 182
  Rhus radicans
Poison Oak, p. 182
  Rhus toxicodendron
Poison Sumac, p. 186
  Rhus vernix

Spurge-nettle, p. 32
  Cnidoscolus stimulosus
Trumpet-creeper
  Campsis radicans
Wild Parsnip, p. 66
  Pastinaca sativa
Wood-nettle, p. 150
  Laportea canadensis

Internal Poisons (known fatalities indicated by *)

American Bittersweet
  Celastrus scandens
* American Yew, p. 164
  Taxus canadensis

Anemone
  Anemone spp.
Apple-of-peru
  Nicandra physalodes

Arnica
Arnica spp.
Arrow Arum, p. 156
Peltandra virqinica
Arrow-grass
Triglochin spp.
Atamasco-lily, pp. 24, 94
Zephyranthes atamasco
*Azalea
Rhododendron spp.
Balsam Apple, Bitter Gourd
Momordica charantia
*Baneberries, p. 44
Actaea spp.
Black Henbane
Hyoscyamus niger
*Black Locust, p. 184
Robinia pseudo-acacia
Bloodroot
Sanguinaria canadensis
Blue Cohosh
*Caulophyllum thalic-
troides*
Blue Flag, p. 130
Iris spp.
Bouncing Bet, p. 34
Saponaria officinalis
Buckthorns, p. 220
Rhamnus spp.
Buttercups, p. 70
Ranunculus spp.
Butterfly-weed, p. 92
Asclepias tuberosa
*Canada Moonseed, pp. 50,
198
Menispermum canadensis
*Castor-bean
Ricinus communis
Celandine
Chelidonium majus
Cestrum, Jessamine
Cestrum spp.
Chinaberry
Melia azedarach
Clematis
Clematis spp.
*Common Tansy, p. 90
Tanacetum vulgare

Corn-cockle
Agrostemma githago
Cowbane
Oxypolis rigidior
Daphne
Daphne mezereum
Death Camas
Zigadenus spp.
Devil's-bit
Chamaelirium luteum
Dicentras
Dicentra spp.
Dogbanes, pp. 48, 110
Apocynum spp.
Elderberry, p. 172
Sambucus spp.
*Ergot, p. 228
Claviceps spp.
*False Hellebore, p. 148
Veratrum viride
Fly-poison, p. 52
*Amianthium muscaetoxi-
cum*
Fool's-parsley, p. 38
Aethusa cynapium
Four-o'clock
Mirabilis spp.
Goat's-rue, pp. 82, 122
Tephrosia virginiana
Golden Club, p. 62
Orontium aquaticum
Golden-seal
Hydrastis canadensis
Ground-cherries, p. 68
Physalis spp.
Holly, p. 192
Ilex spp.
*Horsechestnut, Buckeye,
p. 172
Aesculus spp.
Horsetail,
Equisetum spp.
Horse-nettle
Solanum carolinense
Hydrangea
Hydrangea spp.
Jack-in-the-pulpit, p. 156
Arisaema atrorubens

Jerusalem-oak, p. 152
  Chenopodium botrys
* Jimsonweed, p. 20
  Datura stramonium
Kentucky Coffee-tree, p. 180
  Gymnocladus dioica
* Lantana
  Lantana camara
Larkspur, p. 132
  Delphinium spp.
* Laurel
  Kalmia spp.
Lobelia
  Lobelia spp.
Marsh-marigold, p. 70
  Caltha palustris
May-apple, p. 20
  Podophyllum peltatum
Melonette
  Melothria pendula
Mexican-tea, p. 152
  *Chenopodium ambrosi-
    oides*
* Mistletoe
  Phoradendron flavescens
* Monkshood, p. 132
  Aconitum spp.
Mulberries, p. 210
  Morus spp.
* Mushrooms, p. 238
* Nightshades, pp. 50, 134, 198
  Solanum spp.
* Poison Hemlock, p. 38
  Conium maculatum
* Pokeweed, p. 46
  Phytolacca americana
Prickly Poppy
  Argemone mexicana
Privet
  Ligustrum vulgare
Rattlebox, p. 82
  Crotalaria sagittalis

Rattlebox
  Daubentonia punicea
* Rhododendron
  Rhododendron spp.
Scarlet Pimpernel
  Anagallis arvensis
Scotch Broom, p. 182
  Cytissus scoparius
Skunk Cabbage, p. 156
  Symplocarpus foetidus
Snow-on-the-mountain
  Euphorbia marginata
Spurge
  Euphorbia spp.
Star-of-bethlehem, p. 52
  Ornithogalum umbellatum
Velvet-grass
  Holcus lanatus
Vetchling, Wild Pea
  Lathyrus spp.
Virginia Creeper, p. 180
  *Parthenocissus quinque-
    folia*
* Water-hemlock, p. 42
  Cicuta spp.
* White Snakeroot
  Eupatorium rugosum
Wild Calla, p. 22
  Calla palustris
* Wild Cherries, p. 218
  Prunus spp.
Wild Indigo, p. 80
  Baptisia tinctoria
Wild Lupine, p. 142
  Lupinus perennis
Wisteria, p. 180
  Wisteria spp.
* Yellow Jessamine
  Gelsemium sempervirens

Conservation

SINCE the publication of Rachel Carson's *Silent Spring* in 1962, large numbers of Americans have become increasingly alarmed at the imminent loss of many of our more unique and irreplaceable plants and animals. As a measure of the nation's concern, the Bureau of Endangered Species was established by Congress in 1966 and the U.S. Department of the Interior has recently published a list of threatened or endangered fauna and flora. On a more local level, a variety of conservation groups have set up numerous nature preserves or wildlife refuges, and several states have published lists of protected species. Unfortunately, the majority of the publicity and concern has been directed at the preservation of endangered animals, with the preservation of endangered plants often assuming a secondary role in the public eye. As a result, an increasing number of plants become endangered each year.

Rare and endangered plants fall into one or more of the following three categories: (1) Attractive or useful species that have been actively sought and collected (such as Ginseng, p. 146); (2) species that occupy habitats that are being developed for other uses (such as bog plants); and (3) species that were never plentiful. Today, most plants become threatened or lost through a combination of development and ignorance — development of suitable habitats into shopping centers or parking lots, and ignorance as to the exact nature and value of the plants being destroyed.

In our sophisticated age of automobiles and supermarkets, we have become insulated from the pastoral world of our forebears. Since most of us are no longer directly dependent on the land for our sustenance, we have lost the ability to discern the negative consequences of our acts. A prime intent of many conservation organizations is to bring man into closer contact with the natural environment. Only by doing so will the general populace achieve an understanding of the delicate balance that exists between all living things and learn to respect the plants and animals around them. It may seem to some that a book advocating the collection and use of edible wild plants will only add to man's depredation of his surroundings. I take a different view. Encouraging people to use their powers of observation to find food in the wild will increase their awareness of the natural processes. Actively searching for and then eating a plant can provide some extraordinary insights into the interlocking mosaic of life.

Preservation does not mean that no plant should be picked. It does mean, however, that no plant that is rare or endangered should be picked, and even common plants should be picked in such a way as to insure their survival. The following are some thoughts to keep in mind as you are using this book.

1. A few of the species discussed are relatively uncommon and

should be used only in emergencies (see text on individual species).

2. Some species are common in certain parts of their range and rare in others. These are usually indicated in the text by the words "locally abundant" or "use only when found in abundance."

3. Do not collect more than you will use.

4. Always leave enough for the next person and to insure the plant's survival the next year.

5. When collecting any part other than the root, leave the root in place and intact.

6. When collecting leaves from perennials, do not completely denude the plant; leaves are needed so that the plant can manufacture enough food to survive the winter.

7. When picking a plant, create as little disturbance to the surrounding vegetation as possible.

8. Fragile habitats such as bogs, alpine tundra, and dune communities are particularly susceptible to disturbance and should be entered only infrequently.

Before collecting any plant, obtain, if possible, a list of the threatened or endangered species in your state. Lists can usually be obtained from a state chapter of the American Federation of Garden Clubs. By avoiding the species listed, following the guidelines above, and using your own common sense, you should be able to enjoy edible wild plants without appreciably affecting either their numbers or their surroundings.

Glossary

Most of the words in this *Field Guide* are familiar and need no explanation. However, a few terms, mainly applying to plant parts, require definition. Plant parts not described below are included in the illustrations on the endpapers (front endpaper, roots and flowers; rear endpaper, leaves).

Annual: Living for only a single year (growing season).

Anther: The enlarged, pollen-producing part of the stamen.

Axil: The upper angle between the leaf and the stem it joins.

Biennial: Living for 2 years; usually blooming the second year.

Bract: Modified leaf (often colored) situated near the flower.

Bulb: A short underground stem encased in onionlike scales.

Bulblet: A small bulblike body, especially one on a stem or in a flower cluster. See wild onions, p. 114.

Calyx: The outer whorl of floral leaves (sepals), which may be separate or fused.

Cathartic: A strong purgative.

Colonial: Growing in colonies, usually connected underground.

Composites: Members of the Composite or Daisy Family (sunflowers, dandelions, and so forth).

Compound (leaf): Divided into two or more leaflets. The leaflets can be further subdivided, twice-compound, or even thrice-compound.

Corm: The enlarged base of a stem; bulblike but solid, not layered like an onion.

Deciduous: Foliage shed after each growing season; not evergreen.

Dermatitis: Inflammation of the skin.

Disk: The compact center of some composites (such as sunflowers), composed of many tiny tubular *disk flowers;* commonly encircled by straplike *ray flowers.* See front endpaper.

Emetic: A substance that causes vomiting.

Fiddlehead: The coiled young shoot of a fern.

Frond: A fern leaf; the expanded leaflike part of a seaweed.

Head: A dense cluster of stalkless (or nearly stalkless) flowers.

Herbaceous: Fleshy, non-woody; leaflike in color and texture.

Node: A joint where one or more leaves are attached; any swollen or knoblike structure.

Palm-shaped (palmate): Radiating from a common point (as the fingers of a hand).

Perennial: Living for more than two years.

Petal: One of the segments of the inner floral envelope: usually colored, showy; may be joined basally or separate.

Pistil: The female structure(s) at the center of a flower, from which the fruit develops after fertilization.

Pith: The spongy or hollow core of most stems.

Purgative: A laxative.

Rays, ray flowers: See Disk.

Rootstock (rhizome): An elongate, prostrate or underground stem, usually horizontal and rooting at the nodes.

Runner: A slender trailing shoot that roots at the nodes.

Sepal: One segment of the calyx; usually green, sometimes colored like petals.

Shrub: A woody plant usually no more than 15 ft. in height and with several stems or trunks from the base.

Spadix: A fleshy, club-shaped or fingerlike stalk crowded with tiny flowers (the "Jack" or preacher in the canopied pulpit of Jack-in-the-pulpit, p. 156).

Spathe: A modified leaf (bract) partially enclosing a spadix, or at the base of a flower cluster (see wild onions, p. 114).

Spur: A saclike or tubular extension on a flower.

Stamen: The male flower organ (usually several) consisting of a knoblike, pollen-bearing *anther* atop a slender stem.

Stigma: The usually sticky, knobbed or divided tip of the pistil to which pollen adheres during fertilization.

Stipule: A small leaflike appendage at the base of a leafstalk.

Taproot: The primary root continuing the axis of the plant downward (as in a parsnip or carrot).

Tuber: A thickened, short, underground branch with numerous buds or eyes (as in a potato).

Weed: A plant that grows where it is not wanted, especially one that crowds out more desirable species in cultivated areas.

Whorl: Three or more leaves, or other plant part, radiating from a common point (like the spokes of a wheel).

Wing: A flattened, fleshy or corky membrane projecting from a stem, stalk, fruit, or seed.

Explanation of Symbols

SYMBOLS are employed in the margins of the text to point out food uses and warn against poisonous plants. Definitions are given below. Methods of preparation for each food type are discussed in the *Food Uses* section (page numbers follow the definitions). Typical example(s) of the plants involved are also listed below.

 ASPARAGUS. Steamed or boiled young shoots served like asparagus. See p. 286. Japanese Knotweed (p. 46); Greenbriers (p. 196).

CANDY. Roots, stems, petals, or nuts simmered in sugar syrup or sprinkled with granulated sugar. See p. 287. Wild Ginger (pp. 96, 160); Violets (p. 132). Also, unsweetened chewing gums. See p. 288. Spruces (p. 168).

CEREAL. Small seeds used as cereals or boiled into mush. See p. 288. Wild Rice (p. 228).

COFFEE. Fruit or roots slow-roasted and ground, and prepared like coffee or hot chocolate. See p. 289. Cleavers (p. 50); Chicory (p. 144).

COLD DRINK. Chilled and sweetened fruit juices or plant extracts. See p. 290. Cranberries (pp. 102, 222); Sumacs (p. 186).

 COOKED GREEN. Steamed or boiled leaves served like spinach or cabbage. See p. 291. Marsh-marigold (p. 70); Nettles (p. 150); Lamb's-quarters (p. 152). A few can be used okralike to thicken soups and stews. Marsh Mallow (p. 108).

 COOKED VEGETABLE. Boiled, baked, or fried vegetables yielding dishes that resemble parsnips, peas, green beans, celery, or broccoli. Also, mushrooms. See p. 293. Day-lily (p. 92); Beach Pea (pp. 122, 142).

FLOUR. Roots, tubers, seeds, nuts, bark, or pollen that yield flour. See p. 295. Cattails (p. 158); Oaks (p. 204).

FRITTER. Flowers dipped in batter and fried in oil to make fritters. See p. 297. Common Elderberry (pp. 18, 172); Black Locust (p. 184).

FRUIT. Fruit that can be eaten out of hand or cooked to make sauces or stews. See p. 297. Highbush-cranberry (p. 178); Blueberries (p. 220).

JAM/JELLY. Fruit frequently used to make jam and jelly; these may or may not be too tart to be eaten fresh. See p. 299. Crabapples (p. 216); Wild Cherries (p. 218).

15

NUT. Nuts or large seeds that can be eaten out of hand or roasted like chestnuts. See p. 301. Walnuts (p. 188); Oaks (p. 204). A few can also be used to make oil. See p. 301.

PICKLE. Roots, stems, flowerbuds, or young seedpods used to make pickles, capers, and relishes. See p. 302. Purslane (p. 72); Wild Onions (p. 114).

POTATO. Mild-flavored starchy roots, tubers, or corms that can be served like potatoes. See p. 303. Groundnut (pp. 122, 160); Cattails (p. 158).

SALAD. Tender leaves, roots, tubers, shoots, and stems that can be eaten without cooking. See p. 304. Watercress (p. 28); Common Dandelion (p. 84).

SEASONING/CONDIMENT. Fresh or dried leaves, roots, fruit, or seeds used as flavoring agents. Also, grated and moistened roots or seeds used as condiments. See p. 307. Horseradish (p. 26); Bayberries (p. 206).

SYRUP/SUGAR. Tree saps that can be reduced to syrup or sugar when boiled. See p. 309. Maples (p. 176).

TEA. Fresh or dried leaves, roots, flowers, fruit, or bark that make agreeable teas when steeped in hot water. See p. 309. Mints (pp. 54, 118, 138); Sweet Goldenrod (p. 90).

POISONOUS. Plants with sufficient toxicity to cause discomfort, illness, and, in extreme cases, death. See *Poisonous Plants,* pp. 7–10.

Flowering Plants (Mainly Herbaceous)

The divisions between the seven flower colors (white, yellow, orange, pink-red, violet-blue, green, brown) are not always clear-cut. White or whitish flowers may have faint tinges of yellow, pink, or green; yellow-green flowers may often appear green-yellow; and red-purple flowers may merge imperceptibly into violet-blue. Species on the borderline between colors or with more than one color form are repeated.

MISCELLANEOUS FLOWERING SHRUBS

Note: This is an arbitrary selection. See also Woody Plant Section, pp. 163–226.

 JUNEBERRIES, SERVICEBERRIES   **Fruit**
Amelanchier spp.
Shrubs or small trees with drooping clusters of showy 5-petaled flowers that often appear before the oval, toothed leaves. Fruit black, huckleberrylike. Numerous species; typical form shown. **Where found:** Woods, thickets; rocky or swampy areas. Canada south to Ga., Ala. **Flowers:** April–June. **Fruit:** June–Sept.
Use: Fruit, jelly. See p. 220.   SUMMER

LABRADOR-TEA   **Leaves**
Ledum groenlandicum   **Color pl. 4**
A low, evergreen, northern shrub with leathery untoothed leaves. Leaves with *rolled edges;* white or rusty *woolly beneath.* 1–3 ft. (30–90 cm). See also p. 208. **Where found:** Peaty soil, cold bogs. Canada, n. U.S. **Flowers:** May–June.
Use: Tea. The dried leaves make a mild but agreeable tea when steeped in hot water for 5–10 min.   ALL YEAR

COMMON ELDERBERRY   **Flower clusters, ripe**
Sambucus canadensis   **Color pl. 2**   **fruit (only)**
Note the showy flat-topped clusters of tiny 5-petaled flowers followed by juicy purple-black berries. Leaves large, *opposite, compound,* with 5–11 coarsely-toothed leaflets. Twigs soft, stout, with a thick white pith. 3–13 ft. (0.9–3.9 m). See also p. 172. **Where found:** Damp, rich soil; streambanks, thickets, roadside ditches. Manitoba to Nova Scotia, south to Texas, Ga. **Flowers:** June–July. **Fruit:** Aug.–Oct.
Use: Fritters, jelly, cold drink, fruit. See p. 172.
    SUMMER (flowers); LATE SUMMER (fruit)

NEW JERSEY TEA   **Leaves**
Ceanothus americanus   **Color pl. 13**
A low bushy shrub. Note the long-stalked oval clusters of tiny 5-petaled flowers in the upper leaf axils. Leaves finely-toothed, 2–4 in. (5–10 cm) long, with 3 prominent veins curving to the pointed tip. To 4 ft. (1.2 m). See also p. 192. **Where found:** Dry open woods, dry roadbanks. S. Manitoba, s. Quebec, cen. Maine, south to Texas, Fla. **Flowers:** May–Sept.
Use: Tea. Prepare the thoroughly dried leaves as you would oriental tea. Excellent.   SUMMER

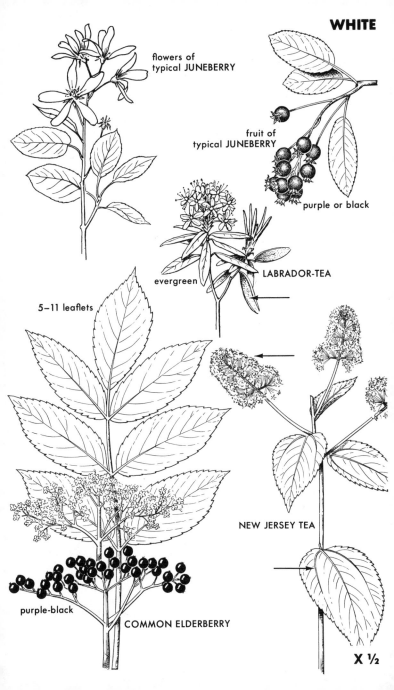

WHITE

flowers of typical JUNEBERRY

fruit of typical JUNEBERRY

purple or black

LABRADOR-TEA

evergreen

5–11 leaflets

NEW JERSEY TEA

purple-black

COMMON ELDERBERRY

X ½

MISCELLANEOUS SHOWY FLOWERS

MAY-APPLE, MANDRAKE   **Mature Fruit (only)**
Podophyllum peltatum
A single, waxy, 6- to 9-petaled flower nods in the fork of the stem below 2 large, deeply cut, umbrellalike leaves. Fruit a large, egg-shaped, greenish yellow berry. Often colonial. 12–18 in. (30–45 cm). **Where found:** Moist woods, openings. Minn. to w. Quebec, south to Texas, Fla. **Flowers:** March–June.
Use: Fresh fruit, jelly, cold drink. The large pale yellow berries ripen as the plants begin to wither and die. The pulp surrounding the seeds can be eaten raw, or cooked and made into jelly (add pectin); the juice can be added to lemonade. **Warning:** The roots, leaves, seeds, and green fruit are strongly cathartic and should not be eaten. Late Summer (Aug.–early Sept.)

WILD POTATO-VINE   **Roots**
Ipomoea pandurata
A trailing vine with singly attached *heart-shaped* leaves and large — 2–4 in. (5–10 cm) — bell-like white flowers with pink-purple centers. Root large, vertical, deeply buried. **Where found:** Dry soil, fields, roadsides, fence rows. Kans., s. Mich., Conn., south to Texas, Fla. **Flowers:** May–Sept.
Use: Potato. Baked or boiled, the immense roots resemble slightly bitter sweet potatoes. Some roots are more bitter than others and should be boiled in several changes of water. **Warning:** The raw roots are a purgative.
Fall–Early Spring

JIMSONWEED, THORNAPPLE   **Poisonous**
Datura stramonium
Note the large — 3–5 in. (7.5–12.5 cm) — trumpet-shaped white or pale violet flowers of this coarse, ill-scented weed. Leaves coarse-toothed; seedpods spiny. 2–5 ft. (0.6–1.5 m). **Where found:** Waste places. Most of our area. **Flowers:** June–Sept.
Warning: All parts of this plant are extremely poisonous.

YUCCA, BEAR-GRASS   **Petals**
Yucca filamentosa
See Yuccas, p. 170.   Summer

BUNCHBERRY, DWARF CORNEL   **Fruit**
Cornus canadensis   **Color pl. 15**
The single showy "blossom" consists of *4 petal-like bracts* around a cluster of tiny flowers. Leaves in a whorl of 6. Berries scarlet, tightly clustered. 3–8 in. (7.5–20 cm). **Where found:** Cold woods, mt. slopes. Canada, northern edge of U.S.; in mts. to W. Va. **Flowers:** May–July. **Fruit:** Late July–Oct.
Use: Fruit. The insipid-tasting ripe berries can be eaten raw or cooked like pudding.   Late Summer–Early Fall

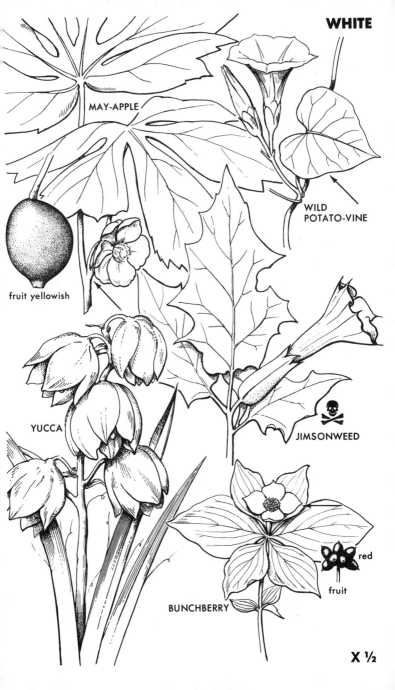

WHITE

MAY-APPLE

WILD
POTATO-VINE

fruit yellowish

YUCCA

JIMSONWEED

BUNCHBERRY

red

fruit

X ½

MISCELLANEOUS AQUATIC FLOWERS

WATER-LILIES   Young leaves, flowerbuds, seeds;
Nymphaea spp.   tubers (Tuberous Water-lily)
Note the platterlike leaves and large showy blossoms floating
on the water's surface. One species shown.
Use: Cooked green, cooked vegetable, flour, potato (Tuberous
Water-lily). The young unrolling *leaves* and unopened *flowerbuds* can be boiled for 5–10 min. and served with butter. The
seeds are rich in starch, oil, and protein, and can be prepared
like those of the Yellow Pond-lilies, p. 60. The rootstock of
Tuberous Water-lily, *N. tuberosa,* produces brown *tubers* the
size of hens' eggs; these can be freed from the mud using your
feet and collected as they float to the surface. Prepare like
potatoes.   SPRING (leaves); SUMMER (flowerbuds);
    FALL (seeds); FALL–EARLY SPRING (tubers)
FRAGRANT WATER-LILY, *N. odorata.* Leaves *purplish*
beneath. Petals tapering to tip. **Where found:** Ponds, quiet
water. Most of our area. **Flowers:** June–Sept.
TUBEROUS WATER-LILY, *N. tuberosa* (not shown). Similar, but leaves usually green beneath. Petals broadly rounded
at tip. **Where found:** Ponds, quiet water. Neb., Minn.,
n. Ontario, s.w. Quebec, south to Ark., Ill., Ohio, Md. **Flowers:**
June–Sept.

BUCKBEAN   Rootstock
Menyanthes trifoliata
The erect flowerstalk and 3 oval leaflets emerge from shallow
water.   Flowers 5-petaled, frosty-white.   Petals *bearded.*
Where found: Bogs, pond edges. N. Canada south to Ill., Ind.,
Ohio, W. Va., and Md. **Flowers:** April–July.
Use: Cooked vegetable, flour. Boil the long, bitter rootstocks
for 30 min. in several changes of water. Or dry, crush, and leach
thoroughly to make an unpalatable but nutritious flour. **Note:**
Use only in extreme need; protected in U.S.
    FALL–EARLY SPRING

WILD CALLA or WATER-ARUM   Dried seeds,
Calla palustris   Color pl. 5   rootstock
A low, creeping plant with a *broad white spathe* and stubby
golden spadix topping a solitary flowerstalk. Leaves glossy,
heart-shaped. Fruit a tight cluster of red berries. **Where
found:** Bogs, pond edges, swamps. Cen. Canada south to
Minn., Ind., Penn., and N.J. **Flowers:** May–Aug.
Use: Flour. The thoroughly dried *seeds* and *rootstocks* can be
ground into an unpalatable but nutritious flour in times of
need.
Warning: Do not eat raw; has the same acrid, biting quality as
Jack-in-the-pulpit, p. 156.   FALL (seeds);
    FALL–EARLY SPRING (rootstock)

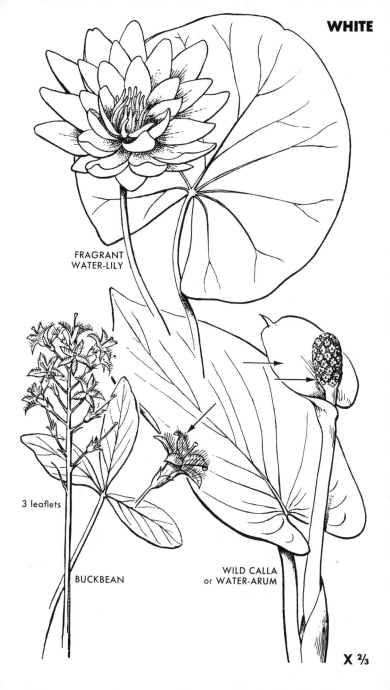

WHITE

FRAGRANT
WATER-LILY

3 leaflets

BUCKBEAN

WILD CALLA
or WATER-ARUM

X ⅔

SHOWY 3- OR 6-PART FLOWERS

ATAMASCO-LILY   **Poisonous**
Zephyranthes atamasco
Note the long, narrow, daffodil-like leaves and erect, waxy, 6-petaled flowers that are usually white but often turn pink with age. Bulb onionlike, but without odor. 6–15 in. (15–37.5 cm). **Where found:** Wet woods, clearings. Va. south to Miss., Fla. **Flowers:** April–June.
Warning: The leaves are quite toxic; the bulbs are suspect

LARGE-FLOWERED or WHITE TRILLIUM   **Very**
Trillium grandiflorum   **young leaves**
Note the whorl of 3 broad leaves and 3 showy petals typical of the trilliums (see also p. 96). Flowers white, turning pink with age. 12–18 in. (30–45 cm). **Where found:** Rich moist woods. Minn. to s. Ontario and w. New England, south to Ark. and Ga. **Flowers:** April–June.
Use: Salad, cooked green. The young unfolding leaves are excellent added fresh to salads, suggesting raw sunflower seeds, or boiled for 10 min. and served with vinegar; the leaves become bitter once the buds and flowers appear. **Note:** Although most trilliums should not be picked, this species and those discussed on p. 96 are occasionally found in sufficient quantity to warrant collection.   EARLY SPRING

ARROWHEADS, DUCK-POTATOES   **Tubers**
Sagittaria spp.
Widespread aquatic plants. Note the *3 roundish petals* and the flowers arranged in *whorls of 3*. Leaves variable, broadly arrowhead-shaped to lancelike or grasslike (sometimes within the same species). Small — 1–2 in. (2.5–5 cm) — potatolike tubers form at the ends of long subterranean runners that originate at the base of each plant. Numerous species in our area.
Use: Potato. The tubers can be gathered in quantity by freeing them from the mud with a hoe or rake and collecting them as they float to the water's surface. Although slightly unpleasant-tasting eaten raw, the tubers are delicious when cooked; prepare them as you would potatoes.   FALL–EARLY SPRING
GRASS-LEAVED ARROWHEAD, *S. graminea.* Leaves lancelike or grasslike. **Where found:** Wet mud or sand, shallow water. Canada south to Texas, La., Ala., and Va. **Flowers:** May–Sept.
SESSILE-FRUITED ARROWHEAD, *S. rigida.* Pistillate flowers and fruit *stalkless.* Leaves occasionally with narrow appendages (see dotted lines). **Where found:** Pond edges, quiet water. Minn., Quebec, Maine, south to Mo., Va. **Flowers:** May–Oct.
BROAD-LEAVED ARROWHEAD, WAPATO, *S. latifolia.* Our commonest species. Leaves extremely variable. **Where found:** Pond edges, shallow water. Most of our area. **Flowers:** June–Oct.

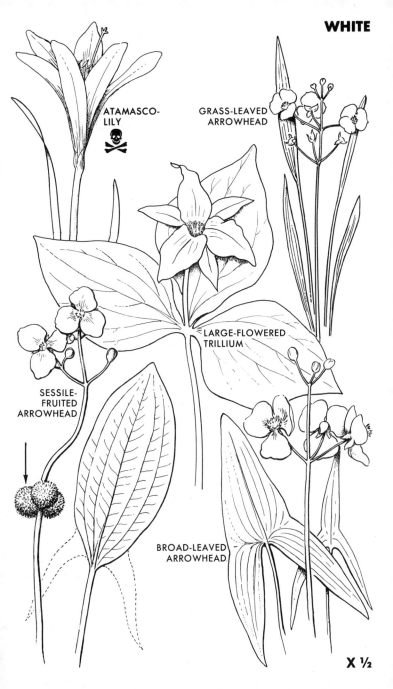

WHITE

ATAMASCO-LILY ☠

GRASS-LEAVED ARROWHEAD

LARGE-FLOWERED TRILLIUM

SESSILE-FRUITED ARROWHEAD

BROAD-LEAVED ARROWHEAD

X ½

4 PETALS; SLENDER CLUSTERS; SHORT PODS

PEPPERGRASSES **Young leaves, green seedpods**
Lepidium spp.
Weeds of roadsides and waste places that form low rosettes of toothed or deeply cut basal leaves in spring and crowded spike-like clusters of tiny 4-petaled flowers soon after. Seed-pods flat, circular or nearly so, slightly notched. Seeds (1 in each half of the pod) peppery. 2 species shown.
Use: Salad, cooked green, seasoning. Add the pungent *young leaves* to salads or boil for 10 min. Add the peppery *green seedpods* to hot soups and stews. Leaves contain vitamins C and A, iron, and protein.   SPRING (leaves); SUMMER (pods)
COW-CRESS, FIELD PEPPERGRASS, *L. campestre.*
Basal lobes of leaves *embrace stem.* 8–18 in. (20–45 cm).
Where found: Waste places, fields. Throughout. **Flowers:** May–Sept.
POOR-MAN'S-PEPPER, PEPPERGRASS, *L. virginicum.*
Similar to Cow-cress, but leaves more deeply toothed, *stalked.* 6–24 in. (15–60 cm). **Where found:** Dry soil, waste ground. Throughout, but rare in far North. **Flowers:** June–Nov.

SHEPHERD'S PURSE **Young leaves, seedpods**
Capsella bursa-pastoris
Somewhat similar to the peppergrasses (above), but the flat seedpods are *heart-shaped.* Basal leaves *dandelionlike;* stem leaves clasping. 8–20 in. (20–50 cm). **Where found:** Roadsides, waste ground. Throughout. **Flowers:** March–Dec.
Use: Salad, cooked green, seasoning. Add the *young leaves* to salads, or prepare like spinach; gather before the flowers appear. Use the dried *seedpods* as a pepperlike seasoning.
SPRING–SUMMER (leaves); LATE SUMMER–FALL (pods)

HORSERADISH **Young leaves, roots**
Armoracea lapathifolia
A tall, coarse plant. Note the large, round-toothed, *long-stalked basal leaves* and *tiny egg-shaped seedpods.* Roots large, white, peppery. 2–4 ft. (0.6–1.2 m). **Where found:** Moist waste ground. Throughout. **Flowers:** May–July.
Use: Condiment, salad. The source of commercial horseradish. Mix the grated *roots* with a little vinegar. Add the tender *young leaves* to salads.   SPRING (leaves); ALL YEAR (roots)

FIELD PENNYCRESS **Young leaves, seedpods**
Thlaspi arvense
Similar to Shepherd's Purse (above), but note the large, circular, *deeply notched* seedpods ("pennies"). Stem leaves wider; lacks dandelionlike basal leaves. 6–18 in. (15–45 cm). **Where found:** Waste ground, fields. Throughout, but commonest in North. **Flowers:** April–Aug.
Use: Salad, cooked green, seasoning. See Shepherd's Purse (above).   SPRING–SUMMER (leaves);
LATE SUMMER–FALL (pods)

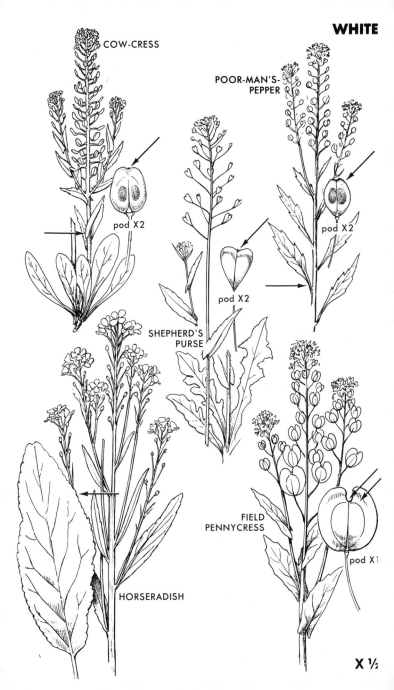

WHITE

COW-CRESS

pod X2

POOR-MAN'S-PEPPER

pod X2

SHEPHERD'S PURSE

pod X2

HORSERADISH

FIELD PENNYCRESS

pod X1

X ½

4 PETALS, SMALL CLUSTERS;
MOIST OR WET PLACES

TOOTHWORTS   **Rootstock**
Dentaria spp.   **Color pl. 14**
Usually white, but fading to pink. See p. 100.   Spring

WATERCRESS   **Young leaves and stems**
Nasturtium officinale
Floating or creeping; often forming dense mats. Leaves with
3–9 small oval leaflets; succulent, pungent tasting. 4–10 in.
(10–25 cm). **Where found:** Running water, springs, brooks.
S. Canada south. **Flowers:** April–Oct.
Use: Salad. Outstanding.   All Year

SPRING CRESS   **Young basal leaves, rootstock**
Cardamine bulbosa
A slender, erect flower of early spring. Basal leaves *long-
stalked, roundish;* stem leaves stalkless, toothed. Rootstock
bulbous, crisp, white. Leaves and rootstock peppery-pungent.
8–20 in. (20–50 cm). **Where found:** Springs, low-lying woods,
wet meadows. Minn. to N.H. and south. **Flowers:** March–
June.
Use: Salad, condiment. Add the tender *young basal leaves* to
salads. Add vinegar to the finely chopped young leaves or
grated *rootstocks* to make horseradish.   Early Spring

MOUNTAIN WATERCRESS   **Young leaves and stems**
Cardamine rotundifolia
A weak, fleshy plant that often forms mats. Leaves roundish;
lower leaves with small projections on leafstalks. **Where
found:** Cold springs, brooks, wet spots. Ohio, w. N.Y., Penn.,
N.J., south in mts. to Ky., N.C. **Flowers:** April–June.
Use: Salad. Delicious; mildly pungent.   Spring

CUCKOO-FLOWER, LADY'S-SMOCK   **Young basal**
Cardamine pratensis   **leaves**
Note the *numerous small paired leaflets;* roundish on basal
leaves, narrow on upper leaves. Basal leaves in rosettes. Flow-
ers white or pale rose. 8–20 in. (20–50 cm). **Where found:**
Swamps, springs, wet meadows, wet woods. N. Canada south to
Minn., n. Ill., n. Ohio, n. N.J. **Flowers:** April–June.
Use: Salad. Delicious; mildly pungent.   Early Spring

PENNSYLVANIA BITTERCRESS   **Young basal leaves**
Cardamine pensylvanica
Resembles Cuckoo-flower (above) but *flowers tiny.* 8–24 in.
(20–60 cm). **Where found:** Springs, brooks, wet ground. Most
of our area. **Flowers:** March–July.
Use: Salad. Our native watercress.   Spring

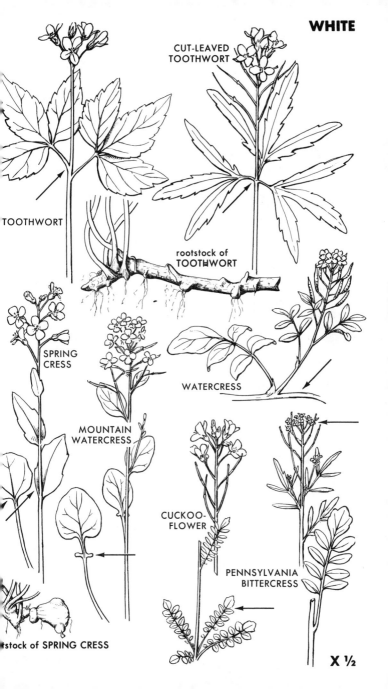

WHITE

CUT-LEAVED TOOTHWORT

TOOTHWORT

rootstock of TOOTHWORT

SPRING CRESS

WATERCRESS

MOUNTAIN WATERCRESS

CUCKOO-FLOWER

PENNSYLVANIA BITTERCRESS

stock of SPRING CRESS

X ½

5 PETALS; MOSTLY 3 LEAFLETS

COMMON WOOD-SORREL   **Leaves**
Oxalis montana   **Color pl. 15**

A low, delicate, woodland flower. Leaves *cloverlike* with 3 inversely heart-shaped leaflets that often fold along a central crease. *Note sour taste.* Petals white or pale pink with prominent pink veins. Colonial. 3–4 in. (7.5–10 cm). See also Yellow Wood-sorrel, p. 72, and Violet Wood-sorrel, p. 104. **Where found:** Cool, moist, deciduous or evergreen woods. Canada, n. U.S.; in mts. to Tenn., N.C. **Flowers:** May–July.
Use: Salad, cold drink. The fresh leaves are an excellent sour addition to salads. To make a refreshing drink, steep the leaves for 10 min. in hot water, chill, and sweeten. Rich in vitamin C.
Warning: Excessive consumption over an extended period of time may inhibit the absorption of calcium by the body.

                                     SPRING–SUMMER

WILD STRAWBERRIES   **Fruit, leaves**
Fragaria spp.

Low plants similar to cultivated strawberries, but with smaller fruit. Leaves long-stalked, with 3 coarsely-toothed leaflets. Flowers round-petaled, in flat clusters on a separate stalk from leaves. Colonial. 2 species shown.
Use: Fresh or cooked fruit, jam, tea. Although the *fruit* is smaller, wild strawberries are much tastier than domestic varieties. Use like cultivated strawberries; pectin needed when making jam. The dried *leaves* make a pleasant tea. An extract of the fresh leaves is rich in vitamin C.   SUMMER
WOOD STRAWBERRY, *F. vesca.* Not as common as the following species. Flowers and fruit smaller, on stalks that usually rise *above* the leaves. Fruit more conical; seeds *on surface.* 3–6 in. (7.5–15 cm). **Where found:** Moist, rocky woods; openings. Canada, n. U.S. south to Mo., Va. **Flowers:** May–Aug.
COMMON STRAWBERRY, *F. virginiana.* **Color pl. 10.** Hairy. Stalks with flowers and fruit do not rise above leaves. Fruit ovoid; *seeds embedded in pits.* 3–6 in. (7.5–15 cm). **Where found:** Fields, open places. Most of our area. **Flowers:** April–June.

BRAMBLES   **Young shoots (blackberries); leaves, fruit**
Rubus spp.   **Color pl. 11 (Common Blackberry)**
Raspberries, dewberries, and blackberries form a complex group of prickly or bristly shrubs with 3–5 leaflets. See p. 184.
Use: Fruit, jelly, cold drink, tea, salad.
          SPRING (blackberry shoots); SUMMER (leaves, fruit)

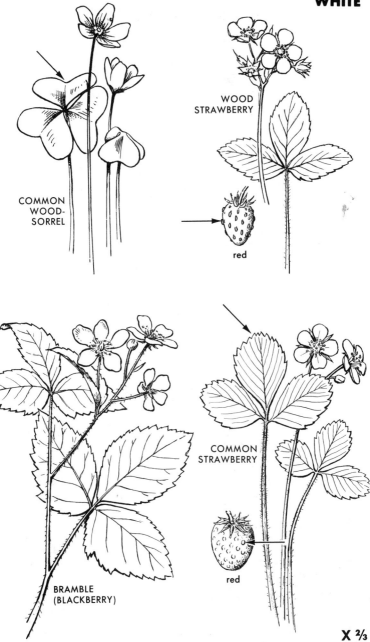

COMMON
WOOD-
SORREL

WOOD
STRAWBERRY

red

BRAMBLE
(BLACKBERRY)

COMMON
STRAWBERRY

red

X ⅔

SPRING-BEAUTIES   **Corm**
Claytonia spp.
A single pair of smooth leaves are attached halfway up the delicate stem. Petals white to pale pink, with darker pink veins. Root a small — $1/4$-$1\frac{1}{2}$ in. (0.6-3.8 cm) — corm; buried 3-5 in. (7.5-13 cm). Colonial. 6-12 in. (15-30 cm) tall.
Use: Potato. Excellent, but tedious to gather in quantity. Boil for 10-15 min., strip off the tough outer jackets, and serve with butter. **Note:** Collect only when abundant.   Spring
CAROLINA SPRING-BEAUTY, *C. caroliniana*. **Color pl. 14.** Leaves *broad,* slender-stalked. **Where found:** Moist woods. S. Canada south through Appalachians and westward. **Flowers:** March–May.
SPRING-BEAUTY, *C. virginica*. Leaves narrow, *linear*. **Where found:** Moist woods, rich soil. Minn. to s. Quebec, south to Texas, Ala., and Ga. **Flowers:** March–May.

SPURGE-NETTLE, TREAD-SOFTLY   **Tuber**
Cnidoscolus stimulosus
Do not touch any aboveground parts (even flowers); armed with *stinging hairs*. Stem stout. Leaves alternate, palm-shaped, with 3-5 irregularly toothed lobes. Flowers in small flat clusters; tubular, with 5 flaring lobes. Stem often continues far underground; root tuberous, sausagelike or irregular, white and starchy within. 6-24 in. (15-60 cm). **Where found:** Dry sandy woods, old fields, roadsides, dunes. Va. south to Fla. and Texas. **Flowers:** Late March–Sept.
Use: Potato. Excellent. Tubers easiest to gather through side excavation of sand dunes. **Warning:** Stinging hairs more virulent than those of nettles, p. 150.   All Year

CHEESES, COMMON MALLOW   **Young leaves, green**
Malva neglecta   **Color pl. 8**   **fruit**
A low creeping or trailing weed. Flowers in leaf axils, pale rose-lavender or white with magenta veins; petals notched. Leaves roundish, with 5-7 scalloped lobes. Fruit *flat, round;* cheeselike in form. **Where found:** Barnyards, waste places. Throughout. **Flowers:** April–Oct.
Use: Okralike thickener, cooked green, salad. See mallows, p. 108.   Spring (leaves); Summer (leaves, fruit)

CLOUDBERRY, BAKE-APPLE   **Fruit**
Rubus chamaemorus   **Color pl. 4**
A small, herbaceous, far-northern "blackberry." A *solitary* flower tops an unbranched stem with 2 or 3 leaves. Leaves broad, 5-lobed. Fruit peach color to amber-yellow. 3-8 in. (7.5-20 cm). **Where found:** Bogs, tundra. Arctic south to n. N.H., Maine. **Flowers:** June–July. **Fruit:** July–Aug.
Use: Fruit, jelly. Delicious cold, with cream and sugar.
                                                  Summer

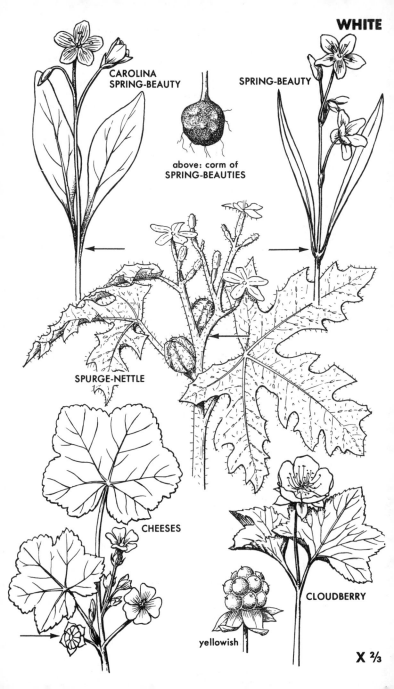

WHITE

CAROLINA
SPRING-BEAUTY

SPRING-BEAUTY

above: corm of
SPRING-BEAUTIES

SPURGE-NETTLE

CHEESES

CLOUDBERRY

yellowish

X ⅔

5 NOTCHED PETALS; LEAVES IN PAIRS

CHICKWEEDS   **Tender leaves and stems**
Stellaria spp. and *Cerastium* spp.

Small, prostrate or erect weeds. Flowers small, mostly long-stalked; petals often so deeply notched or cleft that there appear to be 10. Stems slender. Leaves paired; generally smooth in *Stellaria* spp., downy in *Cerastium* spp. 3 shown. **Use:** Salad, cooked green. The tender leaves and stems can be added to salads, but are best boiled for 5 min. and served as greens. Mouse-ear Chickweed's hairy leaves should always be cooked.   EARLY SPRING–FALL

STAR CHICKWEED, *Stellaria pubera.* Leaves mostly stalkless, broad; up to 3 in. (7.5 cm) long. Petals longer than sepals, cleft ½ or more. 6–16 in. (15–40 cm). **Where found:** Rich, moist soil, woods. Ill. to N.J., south to Ala., n. Fla. **Flowers:** Late March–June.

COMMON CHICKWEED, *Stellaria media.* Our commonest species. Leaves ovate, *long-stalked.* Petals *shorter* than sepals, 2-part. 4–16 in. (10–40 cm). **Where found:** Moist soil, waste places, gardens, roadsides. Throughout. **Flowers:** Most of year.

MOUSE-EAR CHICKWEED, *Cerastium vulgatum.* Leaves *oval, stalkless, hairy.* Stems sticky-hairy. Petals as long as sepals. 6–18 in. (15–45 cm). **Where found:** Roadsides, fields, waste places. Throughout. **Flowers:** April–Sept.

BOUNCING BET, SOAPWORT   **Poisonous**
Saponaria officinalis

Flowers in showy clusters; usually pale pink, but also whitish. Petals scallop-tipped; *reflexed,* sometimes double. Calyx tubular. Stem and leaves smooth; stem thick-jointed. 1–2 ft. (30–60 cm). **Where found:** Roadsides, railroad banks, waste places. Throughout. **Flowers:** May–Oct.
Warning: Saponins contained in this plant can cause severe irritation to the digestive tract. The crushed green plant can be used as a soap substitute.

BLADDER CAMPION   **Young leaves**
Silene cucubalus

The *balloonlike, veined calyx sac* resembles a tiny melon. Petals deeply cleft. Stem and leaves smooth, blue-green (dusted with white powder); leaves stalkless. 8–18 in. (20–45 cm). **Where found:** Dry soil, roadsides, borders of fields, waste places. Canada south to Kans., Mo., Tenn., and Va. **Flowers:** April–Aug.
Use: Cooked green. Collect the tender young leaves when the plant is only a few inches high; boil for 10 min. and serve with butter or vinegar. **Note:** The slight bitterness of the fresh leaves is due to a harmless amount of the toxin saponin.
                                        EARLY SPRING

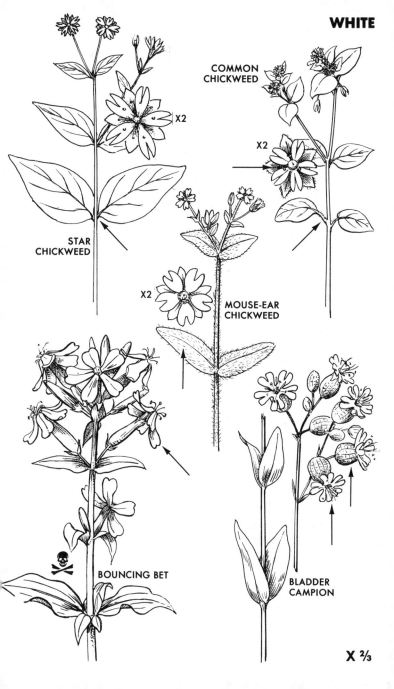

WHITE

COMMON CHICKWEED

X2

STAR CHICKWEED

X2

MOUSE-EAR CHICKWEED

X2

BOUNCING BET

BLADDER CAMPION

X ⅔

CREEPING OR MATTED EVERGREEN PLANTS

WINTERGREEN, CHECKERBERRY **Leaves, fruit**
Gaultheria procumbens
Leaves thick, shiny, oval; *smell of wintergreen*. Small egg-shaped flowers and dry red berries dangle beneath leaves. 2–5 in. (5–12.5 cm). **Where found:** Poor soil, woods. Canada, n. U.S.; south in mts. to Ala., Ga. **Flowers:** July–Aug. **Fruit:** Aug.–next June.
Use: Tea, fruit, salad. See p. 224.   ALL YEAR (leaves); FALL–EARLY SPRING (fruit)

BEARBERRY, KINNIKINIK **Fruit, leaves**
Arctostaphylos uva-ursi
A low, trailing shrub with papery reddish bark. Leaves small, *paddle-shaped*. Flowers pink or white; egg-shaped, with small, lobed mouths. Fruit a red berry. **Where found:** Exposed rock or sand. Arctic south locally to n. U.S. **Flowers:** May–July.
Use: Fruit, tobacco. See p. 224.   SUMMER (leaves); LATE SUMMER–FALL (fruit)

TRAILING ARBUTUS **Corolla (flower tube)**
Epigaea repens
A trailing evergreen shrub with *leathery oval leaves* and woody, brown-hairy stems. Flowers pink or white, clustered, with 5 lobes flaring from a short tube. **Where found:** Dry soil, woods, clearings. Canada, n. U.S.; south in uplands and mts. to Ala., Ga. **Flowers:** Late Feb.–May.
Use: Nibble, salad. The raw corolla, or flower tube, makes an excellent sour-sweet nibble or addition to salads. **Note:** Protected in many states.   EARLY SPRING

PARTRIDGEBERRY **Fruit**
Mitchella repens   **Color pl. 15**
Twin, pink or white, 4-petaled flowers at the end of the creeping stem produce a single red berry. Leaves paired, roundish; may be variegated with whitish lines. **Where found:** Moist woods. Much of our area. **Flowers:** May–July. **Fruit:** July–winter.
Use: Nibble, salad. The dry, seedy berries make a colorful, if tasteless, addition to salads.   FALL–EARLY SPRING

CREEPING SNOWBERRY, MOXIE-PLUM **Leaves,**
Gaultheria hispidula   **Color pl. 15** **fruit**
A ground-hugging, creeping shrub. Leaves tiny, alternate, oval; *smell of wintergreen*. Tiny, 4-lobed, bell-like flowers and white berries in leaf axils. **Where found:** Mossy (usually evergreen) woods, bogs, logs. Canada, n. U.S.; south locally in Appalachians. **Flowers:** May–June. **Fruit:** Aug.–Sept.
Use: Tea, fresh or cooked fruit. See p. 222.
ALL YEAR (leaves); LATE SUMMER (fruit)

WINTERGREEN

red

BEARBERRY

red

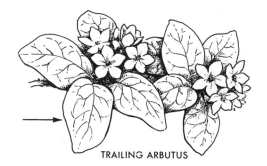

TRAILING ARBUTUS

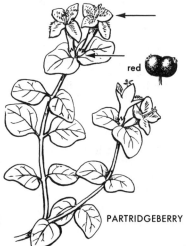

PARTRIDGEBERRY

red

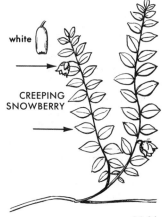

white

CREEPING
SNOWBERRY

X ⅔

UMBRELLALIKE CLUSTERS;
FINELY CUT LEAVES

Caution: Because of the risks involved in a misidentification, beginners should make no attempt to use the species below.

CARAWAY   **Young leaves, seeds, roots**
Carum carvi
A slender hairless biennial with narrowly-cut, olive gray-green leaves. Flowers usually lack bracts. Note the slightly curved, ribbed, aromatic seeds. 1–2 ft. (30–60 cm). **Where found:** Waste places. E. Canada, ne. U.S. **Flowers:** May–July.
Use: Salad, cooked green, seasoning, cooked vegetable. The *young leaves* are excellent in salads or boiled for 10–15 min. The *seeds* are a familiar seasoning. The first-year *roots* can be prepared like parsnips. **Caution:** Young foliage similar to Fool's-parsley and Poison Hemlock (below).
SPRING (leaves); LATE SUMMER–FALL (seeds); FALL–EARLY SPRING (roots)

YARROW   **Leaves**
Achillea millefolium   **Color pl. 7**
Flowerheads with 5 petal-like rays, in flat-topped clusters. Leaves narrow, woolly, aromatic. 1–3 ft. (30–90 cm). **Where found:** Fields, roadsides. Throughout. **Flowers:** June–Sept.
Use: Tea. Steep the dried leaves 10–15 min.   SUMMER

FOOL'S-PARSLEY   **Poisonous**
Aethusa cynapium
Note beardlike bracts below secondary flower clusters. Stems smooth. Foliage ill-scented. 1–2½ ft. (30–75 cm). **Where found:** Waste places. S. Canada, n. U.S. **Flowers:** June–Aug.
Warning: Do not confuse with Wild Carrot (below) or parsley.

WILD CARROT, QUEEN ANNE'S LACE   **Roots**
Daucus carota
A widespread *hairy-stemmed* biennial. Flower clusters flat-topped, lacy; often with a single *purple* flower in center. Old clusters resemble *birds' nests*. Bracts *stiff, 3-forked*. Root white, smells of carrot. 2–3 ft. (60–90 cm). **Where found:** Fields, waste places. Throughout. **Flowers:** May–Oct.
Use: Cooked vegetable. Prepare the first-year roots like garden carrots. **Caution:** Early leaves resemble Poison Hemlock (below), but stalks *hairy*.   FALL–EARLY SPRING

POISON HEMLOCK   **Poisonous**
Conium maculatum
A tall, much-branched biennial. Stems stout, hollow, grooved, *spotted with purple*. Ill-scented when bruised, unpleasant to taste. Root white, carrotlike. 2–6 ft. (0.6–1.8 m). **Where found:** Waste ground. Iowa to Quebec south. **Flowers:** May–Aug.
Warning: Small amounts may cause paralysis and death. Similar to Wild Carrot (above), but leafstalks *hairless*.

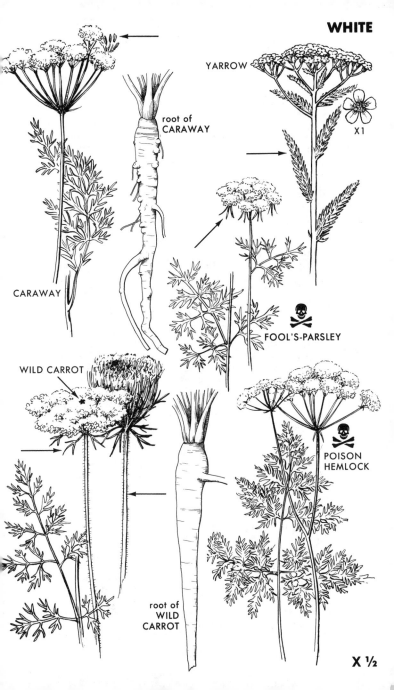

WHITE

YARROW

X1

root of
CARAWAY

CARAWAY

FOOL'S-PARSLEY

WILD CARROT

POISON
HEMLOCK

root of
WILD
CARROT

X ½

UMBRELLALIKE FLOWER CLUSTERS;
BROAD 3- OR 5-PART LEAVES

HONEWORT, WILD CHERVIL   **Young leaves**
Cryptotaenia canadensis   **and stems, roots**
Branching. Leaves long-stemmed, 3-part; leaflets sharply
toothed, often lobed. Flowers tiny, in loose clusters. Note the
slender, ribbed fruit. 1–3 ft. (30–90 cm). **Where found:** Moist
woods. Canada south to Texas, Ga. **Flowers:** May–Sept.
Use: Seasoning, cooked green, cooked vegetable. Add the
young leaves and stems parsleylike to soup or boil for 10–15
min. Prepare the *roots* like parsnips.   SPRING

SPIKENARD, LIFE-OF-MAN   **Roots**
Aralia racemosa
A spreading plant with a smooth dark stem. Leaves up to 3 ft.
(90 cm) long, with 3 major divisions which usually have *15–21
heart-shaped leaflets.* Flowers tiny, in small rounded clusters
arranged along a central stem. Root large, spicy-aromatic. 3–5
ft. (0.9–1.5 m). **Where found:** Rich woods. S. Canada, n. U.S.;
uplands to ne. Kans., Ga. **Flowers:** June–Aug.
Use: Tea, root beer. **Note:** Related, but not shown, WILD
SARSAPARILLA, *A. nudicaulis,* and BRISTLY SARSAPA-
RILLA, *A. hispida,* found in dry woods. They can be used
similarly.   ALL YEAR

COW-PARSNIP   **Young stems and leafstalks,**
Heracleum maximum   **Color pl. 12**   **seeds, roots**
Immense, woolly, rank-smelling. Stems very stout, hollow,
ridged. Flower clusters up to 8 in. (20 cm) across; petals
notched, often tinged with purple. Leaves large — often over 1
ft. (30 cm) — with *3 maplelike leaflets;* leafstalks *swollen* at
base. 4–10 ft. (1.2–3 m). **Where found:** Moist ground, thickets. Canada, n. U.S.; mts. to Ga. **Flowers:** June–Aug.
Use: Cooked vegetable, seasoning. The *young stems and leafstalks* resemble cooked celery when boiled for 15–20 min. in
several changes of water. Tender *roots* can be cooked like
parsnips. Dried *seeds* can be used as a seasoning. **Caution:** Do
not use Water-hemlock (p. 42) by mistake.
                    SPRING (stems, leafstalks); SUMMER (seeds);
                             FALL–EARLY SPRING (roots)

ANGELICA, ALEXANDERS   **Young stems**
Angelica atropurpurea   **Color pl. 6**
Stem smooth, *dark purple.* Upper leaves with swollen basal
sheath. Leaves with 3 leaflets; leaflets *further divided into 3's
or 5's.* 4–9 ft. (1.2–2.7 m). **Where found:** Streambanks,
swamps. Canada south to Ill., Del. **Flowers:** June–Oct.
Use: Cooked vegetable, candy. Peel, and prepare like Cow-
parsnip (above), or make into candy like Wild Ginger (p. 96).
Caution: See Water-hemlock, p. 42.
                             LATE SPRING–EARLY SUMMER

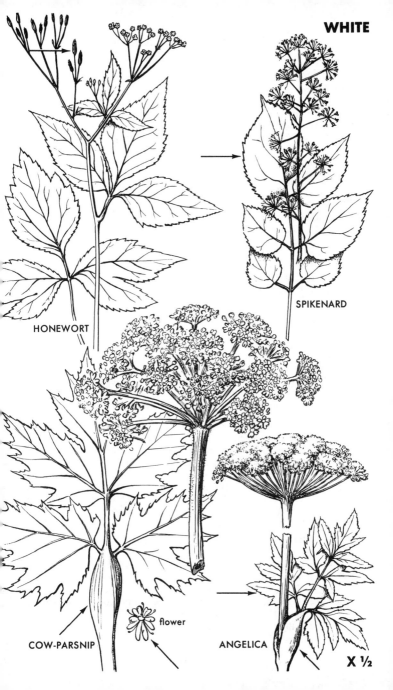

WHITE

SPIKENARD

HONEWORT

COW-PARSNIP

flower

ANGELICA

X ½

UMBRELLALIKE CLUSTERS;
TOOTHED LEAFLETS

 WATER-HEMLOCK, SPOTTED COWBANE Poisonous
Cicuta maculata
Tall, branching, with numerous flower clusters. Stem smooth, *streaked with purple,* chambered. Leaves twice- or thrice-compound, often reddish-tinged. Root with fat tuberlike branches, white. 3–6 ft. (0.9–1.8 m). **Where found:** Wet meadows, swamps. Canada south to Texas, N.C. **Flowers:** June–Sept. **Warning:** Our deadliest species. A single mouthful can kill.

 WATER-PARSNIP Roots
Sium sauve
Similar to Water-hemlock (above), but *stems strongly ridged* and leaves *once-compound* with 3–7 pairs of lance-shaped leaflets. Basal leaves (not shown) very finely cut, often submerged. Roots slender. 2–6 ft. (0.6–1.8 m). **Where found:** Swamps, wet meadows. Most of our area. **Flowers:** July–Sept.
Use: Cooked vegetable. Boil until tender. **Caution:** Because of its close similarity to Water-hemlock (above), Water-parsnip is best ignored as a possible food plant.

FALL–EARLY SPRING

 DWARF GINSENG Tuber
Panax trifolius
Note the whorl of 3 leaves, each with 3 (or 5) stalkless leaflets, and the rounded cluster of tiny flowers. Root a small, round tuber. 4–8 in. (10–20 cm). **Where found:** Moist woods. S. Canada, n. U.S.; in mts. to Ga. **Flowers:** April–June.
Use: Nibble, cooked vegetable. The tiny tubers can be eaten raw, as a trailside nibble, or boiled for 5–10 min. SPRING

SWEET CICELY Roots and green fruit
Osmorhiza claytoni
Soft, hairy. Leaves thrice-compound, bluntly toothed, *fernlike;* lower leaves often over 1 ft. (30 cm) long. Flowers tiny, in sparse clusters. Roots and green fruit *smell of anise.* 1½–3 ft. (45–90 cm). ANISE-ROOT, *O. longistylis* (not shown), similar but nearly hairless. **Where found:** Moist woods. S. Canada south to uplands of Mo., Ala., N.C. **Flowers:** May–June.
Use: Aniselike flavoring. SPRING (roots); SUMMER (fruit)

 HARBINGER-OF-SPRING, PEPPER-AND-SALT
Erigenia bulbosa Tuber
Low, delicate. Leaves 1 or 2, divided into narrow-oblong or lobed segments. Note *leafy bracts* beneath flower clusters. Anthers blackish red, conspicuous. Root a small round tuber. 4–9 in. (10–22.5 cm). **Where found:** Moist deciduous woods. Wisc. to s. Ontario, south to Mo., Ala. **Flowers:** Feb.–May.
Use: Nibble, cooked vegetable. See Dwarf Ginseng (above).

EARLY SPRING

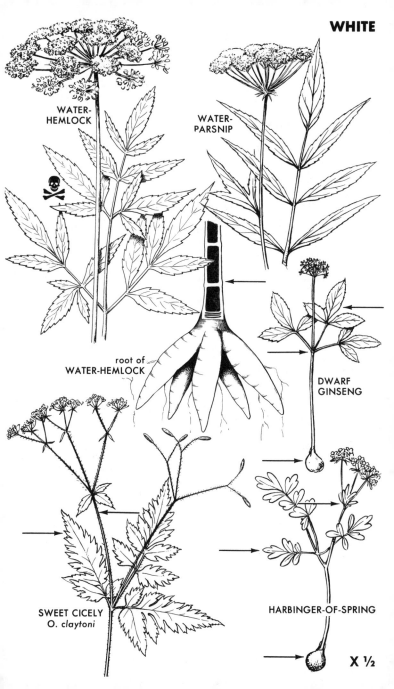

WHITE

WATER-HEMLOCK

WATER-PARSNIP

root of WATER-HEMLOCK

DWARF GINSENG

SWEET CICELY
O. claytoni

HARBINGER-OF-SPRING

X ½

RADIATING CLUSTERS

WATERLEAFS   Young leaves
Hydrophyllum spp.
Note the radiating clusters of white or pale violet, 5-petaled flowers with conspicuously protruding stamens. Leaves often marked with lighter green, as if stained with water. Basal leaves in rosettes. 2 species shown.
Use: Cooked green. The young leaves (before the flowers appear) are excellent boiled for 5-10 min. in 1 or 2 changes of water and served with vinegar.   EARLY SPRING
BROAD-LEAVED WATERLEAF, *H. canadense.* Leaves *maplelike.* Flowerstalks below leaves. 6-20 in. (15-50 cm). **Where found:** Rich moist soil, woods. S. Ontario, w. New England, south to e. Mo., and n. Ala. **Flowers:** June-July.
VIRGINIA WATERLEAF, *H. virginianum.* Leaves irregularly cut, with *5-7 lobes.* Smoothish. 1-3 ft. (30-90 cm). **Where found:** Rich moist soil, woods. Manitoba to Quebec, south to e. Kans., n. Ark., Tenn., Va., and w. New England. **Flowers:** April-Aug.
LARGE-LEAVED WATERLEAF, *H. macrophyllum* (not shown). Similar to Virginia Waterleaf, but *rough-hairy.* Leaves divided into *7 or more lobes.* **Where found:** Rich moist soil, woods. Ill. to W. Va., south to Ala., Ga. **Flowers:** May-June.

BANEBERRIES, DOLL'S-EYES   Poisonous
Actaea spp.
Note the large leaves, divided and subdivided into sharply toothed leaflets. Flowers with very narrow petals (4-10) and bushy stamens; clustered at the end of a long naked stem. Fruit a red or white berry. 2 species shown.
Warning: All parts, particularly the roots and berries, contain a poisonous glycoside. A few berries can cause severe dizziness and vomiting.
WHITE BANEBERRY, *A. pachypoda.* Flower clusters oblong. Berries white, on thick red stalks, tipped with a dark spot ("doll's-eye"). 1-2 ft. (30-60 cm). **Where found:** Woods. S. Canada, south in uplands to Okla., La., Ala., and Ga. **Flowers:** May-June. **Fruit:** July-Oct.
RED BANEBERRY, *A. rubra.* (berries shown). Flower cluster is rounder. Berries red, on thinner stalks. 1-2 ft. (30-60 cm). **Where found:** Woods. S. Canada south to Iowa, Ohio, W. Va., and n. N.J. **Flowers:** May-July. **Fruit:** Aug.-Oct.

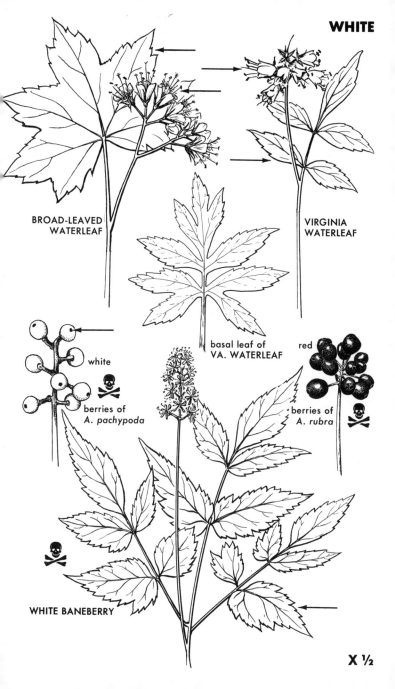

WHITE

BROAD-LEAVED
WATERLEAF

VIRGINIA
WATERLEAF

basal leaf of
VA. WATERLEAF

white

berries of
A. pachypoda

red

berries of
A. rubra

WHITE BANEBERRY

X ½

SLENDER, TAPERING, DENSE CLUSTERS

JAPANESE KNOTWEED   **Young shoots**
Polygonum cuspidatum

A tall, shrublike weed growing in dense, coarse-stemmed stands. Stems hollow, *bamboolike,* with enlarged joints encased in papery sheaths; mottled green, dusted with white powder. Leaves broadly ovate, with pointed tips and squarish bases. Flowers greenish white, profuse; in slender fingerlike clusters in leaf axils. 4–10 ft. (1.2–3 m). **Where found:** Waste places. S. Canada south to Mo., Md. **Flowers:** Aug.–Sept. **Use:** Asparagus, jam. The young shoots — up to 1 ft. (30 cm) — are excellent steamed or boiled for 4–5 min. and served like asparagus; if the flavor is too tart, add a little sugar. Slightly older stems can be peeled and the sour rind boiled with sugar and pectin to make a rhubarblike jam.   SPRING

POKEWEED, POKE   **Young shoots (only)**
Phytolacca americana   **Color pl. 8**

A coarse, widely branched, weedy plant with large leaves and smooth *reddish stems.* Flower clusters long-stalked, often paired with leaves; flowers with 5 greenish white petal-like sepals. Fruiting clusters drooping; berries glossy, purple-black, with red stems, ripening in autumn. 4–10 ft. (1.2–3 m). **Where found:** Roadsides, cultivated fields, waste places. Minn. to Maine and south. **Flowers:** June–Oct.
Use: Asparagus, cooked green, pickle. The young shoots — up to 6 in. (15 cm) — or just the leafy tips, are excellent boiled for 20–30 min. (until tender) in at least 2 changes of water. The peeled shoots can be boiled for 15 min. in several changes of water and pickled in hot vinegar. **Warning:** Root, seeds, and mature stems and leaves dangerously poisonous. Be very careful not to include part of the root when collecting the shoots, and peel or discard any shoots tinged with red.   SPRING

PLANTAINS   **Young leaves**
Plantago spp.

Low, homely plants. Flowers greenish white, tiny; in tight slender heads atop a leafless stem. Leaves in basal rosettes. **Use:** Salad, cooked green. Chop and add to salads, or boil for 10–15 min. and serve with butter. Collect Common Plantain leaves while *very* young; they soon become too stringy to use.
EARLY SPRING (Common P.); SPRING–SUMMER (Seaside P.)
SEASIDE PLANTAIN, *P. juncoides.* A coastal species with narrow (1 rib), *fleshy* leaves. 2–8 in. (5–20 cm). **Where found:** Shores, cliffs. Coast south to N.J. **Flowers:** June–Sept.
COMMON PLANTAIN, *P. major.* The familiar lawn weed. Leaves *broad,* ovate, heavily ribbed, with troughlike stems. Flowerhead *long, very tight.* 6–18 in. (15–45 cm). **Where found:** Lawns. Throughout. **Flowers:** June–Oct.

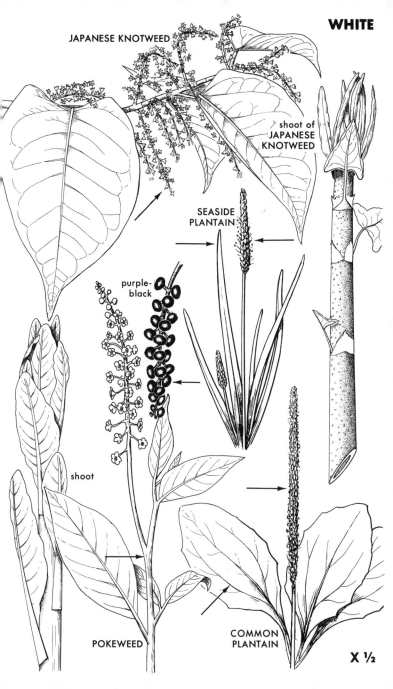

WHITE

JAPANESE KNOTWEED

shoot of
JAPANESE
KNOTWEED

SEASIDE
PLANTAIN

purple-
black

shoot

POKEWEED

COMMON
PLANTAIN

X ½

LOOSE TERMINAL CLUSTERS:
MISCELLANEOUS

SAXIFRAGES   **Young leaves**
Saxifraga spp.
A large group of plants whose principal leaves are usually in
whorls at the base of a slender flowerstalk. Leaves often with
rounded blades and toothed edges. Flowers with 5 petals and
10 bright yellow stamens. 2 species shown.
Use: Salad, cooked green. The young leaves can be added to
salads or boiled for 5–10 min.   EARLY SPRING
LETTUCE SAXIFRAGE, *S. micranthidifolia* (leaf shown).
Similar to Swamp Saxifrage but leaves thinner, *sharply
toothed.* Not as hairy. Flowerstalk more slender; flower cluster
looser. 1–3 ft. (30–90 cm). **Where found:** Brooksides, seeping
banks. Mts. from Penn. to Ga. **Flowers:** April–June.
SWAMP SAXIFRAGE, *S. pensylvanica.* Leaves bluntly
lance-shaped, *toothless* or nearly so; variable. Flowers may be
whitish, greenish, yellowish, or purplish. 1–3½ ft. (0.3–1.1 m).
Where found: Swamps, wet meadows, seeping banks. Minn.
to s. Maine, south to Mo. and Va. **Flowers:** April–June.

BASTARD-TOADFLAX   **Green fruit**
Comandra umbellata
A small, slender-stemmed plant with numerous elliptic leaves.
Leaves attached singly; pale beneath, with a pale midrib.
Flowers small, erect, with a cuplike base and 5 petal-like sepals
(no petals) on the rim. Fruit small — ¼ in. (0.6 cm) — urn-
shaped, crowned by the 5 sepals; green, later drab brown. 6–16
in. (15–40 cm). **Where found:** Dry woods, thickets. Mich. to
Maine, south to n. Ala., Ga. **Flowers:** April–June.
Use: Nibble. The small *green* fruit are a sweet, if not overly
abundant, trailside nibble.   SUMMER

DOGBANES   **Poisonous**
Apocynum spp.
Branching plants with paired, ovate leaves and reddish stems.
Flowers whitish or pale pink; fragrant, bell-like. See p. 110.
Seedpods in pairs, slender, 3–8 in. (7.5–20 cm) long. Stems with
milky juice. 1–4 ft. (0.3–1.2 m).
Warning: Do not confuse the slender shoots with the stouter
ones of Common Milkweed, p. 112. Fortunately, young dog-
bane shoots are hairless and quickly branching.
INTERMEDIATE DOGBANE, *A. medium.* Intermediate
between Spreading Dogbane (p. 110) and Indian Hemp
(below). Flowers showy; lobes not recurved. Leaves wide.
Where found: Fields, roadsides. S. Canada south to Texas,
Mo., Tenn., Va. **Flowers:** June–Aug.
INDIAN HEMP, *A. cannabinum.* Flowers greenish white, not
as showy. Leaves narrower, tapering. **Where found:** Thickets,
fields, sandy shores. S. Canada south. **Flowers:** June–Aug.

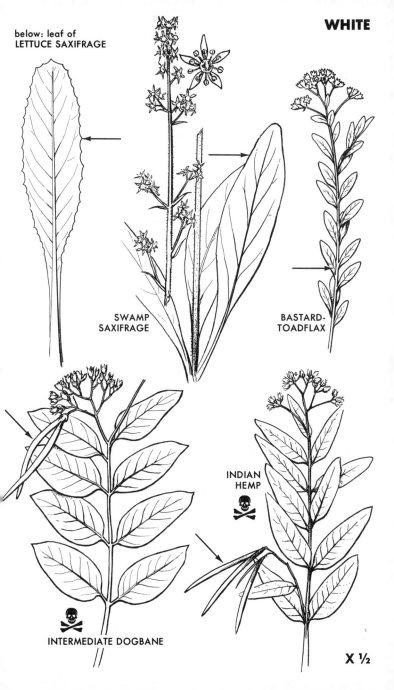

below: leaf of
LETTUCE SAXIFRAGE

WHITE

SWAMP
SAXIFRAGE

BASTARD-
TOADFLAX

INDIAN
HEMP

INTERMEDIATE DOGBANE

X ½

LOOSE CLUSTERS IN AXILS: MISCELLANEOUS

CANADA MOONSEED   **Poisonous**
Menispermum canadense
A woody climbing vine with green twining stems. Leaves
large—5–10 in. (12.5–25 cm)—variable, sometimes nearly round
with a pointed tip, but usually with 3–7 shallow lobes. Leaf
bases *not attached* to leafstalks. Drooping clusters of white
flowers develop into soft grape-sized black fruit powdered with
white. Each fruit has *a single, flattened, crescent-shaped seed.*
Where found: Moist woods, thickets, streambanks.
Se. Manitoba, w. Quebec; w. New England, south to Okla., Ark.,
Ala., and Ga. **Flowers:** May–July. **Fruit:** Sept.–Oct.
Warning: Although this vine strongly suggests a wild grape (p.
198), it lacks tendrils and has only a single crescent-shaped
seed. The bitter, unpalatable fruit are potentially fatal if eaten
in sufficient quantities.

COMMON NIGHTSHADE   **Poisonous**
Solanum nigrum
An erect, bushy weed. Note the 5 reflexed petals and the
beaklike yellow anthers. Leaves long-stalked, irregularly
toothed, roughly triangular. Fruit a black berry. 1–2½ ft.
(30–75 cm). *S. dulcamara* (p. 134) can also be white. **Where
found:** Waste places. Much of our area. **Flowers:** May–Sept.
Warning: Although there is some controversy as to the edibility of the *fully ripe* berries, they should be left alone. The
green berries contain solanine and can be fatal if eaten in
sufficient quantity.

CLEAVERS, GOOSEGRASS   **Young shoots, fruit**
Galium aparine
An inconspicuous, weak-stemmed plant that reclines on surrounding vegetation. Note the scratchy, recurved bristles on
the square stems. Leaves small, narrow, mostly in whorls of 8.
Tiny 4-petaled flowers on slender stalks in leaf axils followed by
tiny, 2-lobed, dry, bristly fruit. **Where found:** Rich moist soil,
thickets, woods, waste ground. Throughout. **Flowers:** May–
July.
Use: Cooked green, salad, coffee. The tender *young shoots* are
excellent boiled for 10–15 min. and served with butter; cooked
shoots can be chilled and added to salads. Slow-roasted until
dark brown and ground, the ripe *fruit* make an outstanding
coffee substitute.   SPRING (shoots); EARLY SUMMER (fruit)

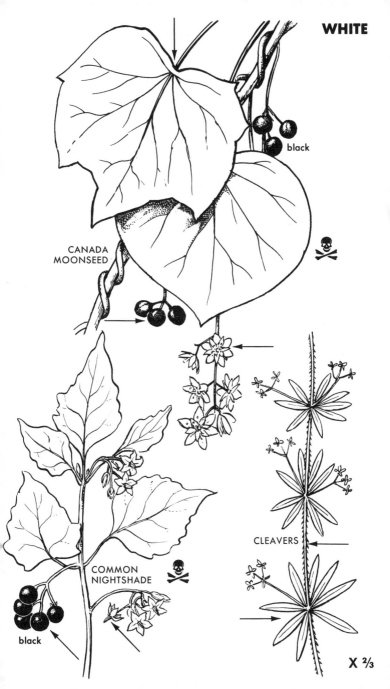

WHITE

black

CANADA
MOONSEED

☠

COMMON
NIGHTSHADE

☠

black

CLEAVERS

X ⅔

6-PART FLOWERS; TERMINAL CLUSTERS; PARALLEL-VEINED LEAVES

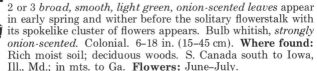

WILD LEEK or RAMP   Leaves, bulb
Allium tricoccum
2 or 3 *broad, smooth, light green, onion-scented leaves* appear in early spring and wither before the solitary flowerstalk with its spokelike cluster of flowers appears. Bulb whitish, *strongly onion-scented.* Colonial. 6–18 in. (15–45 cm). **Where found:** Rich moist soil; deciduous woods. S. Canada south to Iowa, Ill., Md.; in mts. to Ga. **Flowers:** June–July.
Use: Cooked vegetable, pickle, salad, seasoning, cooked green. Our best wild onion. See p. 114.   SPRING (leaves); ALL YEAR (bulb)

FALSE SOLOMON'S-SEAL   Young shoots
Smilacina racemosa
Suggests solomon's-seals (p. 76) with oval, pointed leaves alternating along an arching stem; but flowers creamy-white, in an elongate, frothy, terminal cluster. Fruit a small berry; white speckled with gold at first, ruby-red later. Rootstock with circular "seals" like solomon's-seals, but slender, yellowish. 1½–3 ft. (45–90 cm). **Where found:** Moist woods, clearings, banks. Canada south to uplands of Mo., Tenn., N.C. **Flowers:** April–July.
Use: Salad, asparagus. The *young shoots* can be added to salads or prepared like asparagus. Use only when abundant. **Note:** The *berries* and *rootstocks* are also edible; but the berries are mildly cathartic, and the rootstocks must first be soaked overnight in lye and then parboiled. EARLY SPRING

STAR-OF-BETHLEHEM   Poisonous
Ornithogalum umbellatum
Note the showy, branched cluster of waxy flowers that open only in sunshine; petal with a green stripe underneath. Leaves grasslike, dark green, often with a *whitish midrib.* 4–12 in. (10–30 cm). **Where found:** Fields, roadsides. Neb. to Newfoundland, south to Kans., N.C. **Flowers:** April–June.
Warning: Bulb and leaves contain toxic alkaloids.

FLY-POISON   Poisonous
Amianthium muscaetoxicum
An elongated, fairly dense cluster of round-petaled flowers tops a single smooth stem; creamy-white at first, dull green or pale purple later. Leaves mostly basal; long, narrow, rushlike. Bulb layered like an onion, but lacks the odor. 1–4 ft. (0.3–1.2 m). **Where found:** Moist sandy soil; woods, bogs. S. Mo. to Long Island and south. **Flowers:** May–July.
Warning: Highly toxic; *wash hands* after touching.

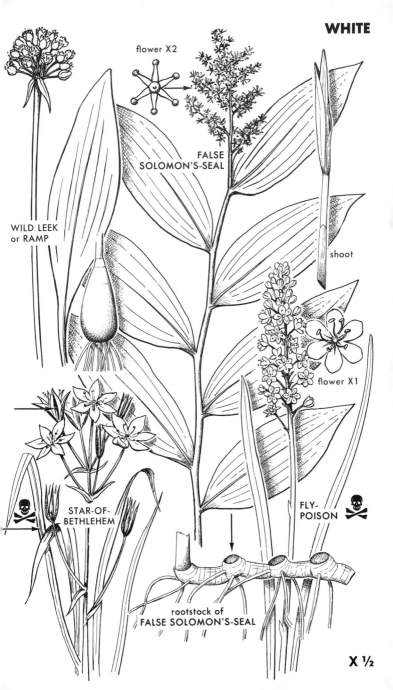

WHITE

flower X2

FALSE
SOLOMON'S-SEAL

shoot

WILD LEEK
or RAMP

flower X1

STAR-OF-
BETHLEHEM

FLY-
POISON

rootstock of
FALSE SOLOMON'S-SEAL

X ½

SQUARE STEMS, PAIRED LEAVES: MINTS
CLUSTERS OF LOBED, 2-LIPPED FLOWERS

WILD MINT Leaves
Mentha arvensis   **Color pl. 2**
White, pale violet, or lavender. See p. 138.   SUMMER

BUGLEWEED Tuber
Lycopus uniflorus
Suggests Wild Mint (above) but odorless. Leaves light green, fine-toothed, short-stalked; narrow, tapering at both ends. Stems slender, *hairless,* rising from a *tuberous* base. Other bugleweeds, *Lycopus* spp., similar but lack tubers. 6–24 in. (15–60 cm). **Where found:** Low, wet ground. Canada south to Neb., the Great Lakes States, and Va. **Flowers:** June–Sept. **Use:** Salad, cooked vegetable, pickle. The crisp tubers are excellent added to salads, pickled, or boiled for 10–15 min. and served with butter.   FALL–EARLY SPRING

CATNIP Leaves
Nepeta cataria
Flowers pale violet or white, purple-spotted. See p. 140.
SUMMER

MOUNTAIN-MINTS Leaves
Pycnanthemum spp.
Widespread, strongly aromatic mints. Most have dense, somewhat flat-topped, branching clusters of compact buttonlike heads on which a few flowers bloom at a time. Flowers small, whitish or purplish, lips often dotted with purple. Many species have conspicuously whitened bracts. Species best identified by leaves; 5 shown. 1–4 ft. (0.3–1.2 m). **Where found:** Dry (or in southern areas, wet) woods, thickets, fields. Most of our area (individual ranges below). **Flowers:** July–Sept.
Use: Tea. Use fresh or dried.   SUMMER
VIRGINIA MT.-MINT, *P. virginianum.* Leaves narrow, toothless; broad at base, tapering to tip. North Dakota to cen. Maine, south to e. Kans., s. New England; in mts. to Tenn., N.C.
SHORT-TOOTHED MT.-MINT, *P. muticum.* Leaves broad, toothed; rounded at base. Mich. to sw. Maine, south to e. Texas, n. Fla.
HAIRY MT.-MINT, *P. pilosum.* Similar to Torrey's Mt.-mint (below) but hairy; leaves broader, stalked. Iowa, s. Mich., and sw. Ontario, south to Okla., Ark., and Tenn.
HOARY MT.-MINT, *P. icanum.* Leaves similar to Short-toothed Mt.-mint (above), but stalked. Undersides covered with white down. S. Ill. to s. N.H., south to La., Fla.
TORREY'S MT.-MINT, *P. verticillatum.* Leaves slender, tapering at base. Mich. to sw. Quebec, south to w. N.C., Va.

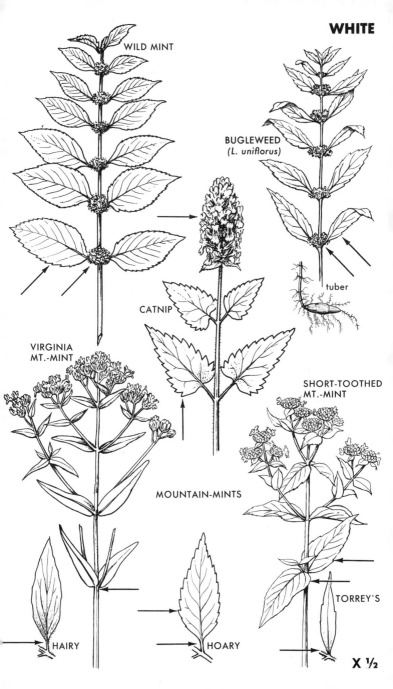

WHITE

WILD MINT

BUGLEWEED
(*L. uniflorus*)

tuber

CATNIP

VIRGINIA
MT.-MINT

SHORT-TOOTHED
MT.-MINT

MOUNTAIN-MINTS

HAIRY

HOARY

TORREY'S

X ½

CLUSTERS OF SMALL PEALIKE FLOWERS;
3 LEAFLETS: CLOVERS

CLOVERS   Young leaves, flowerheads, seeds
Trifolium spp.
Familiar weeds of field and wayside with dense compact heads of tiny pealike flowers and 3 leaflets. Numerous species with white, pink, red, or yellow flowers; 3 white species shown here. See also clovers (p. 124) and Hop Clover (p. 80).
Use: Salad, cooked green, tea, flour. Although not among the choicest of wild foods, clovers are both abundant and rich in protein. The *flowerheads* and tender *young leaves* are difficult to digest raw, but can be eaten in quantity if soaked for several hours in salty water or boiled for 5–10 min. The dried *flowerheads* make a delicate healthful tea when mixed with other teas. The dried *flowerheads* and *seeds* can be ground into a nutritious flour.   SPRING–SUMMER
BUFFALO CLOVER, *T. stoloniferum.* Flowers white or tinged with red. Leaves branch from stem; leaflets *heart-shaped* (notched at tip). Long basal runners. 4–16 in. (10–40 cm). **Where found:** Dry woods, prairies. S. Dak. to W. Va., south to e. Kans., Ky. **Flowers:** May–Aug.
ALSIKE CLOVER, *T. hybridum.* Pale pink or whitish. Leaves unmarked; *branching from stems.* 1–2 ft. (30–60 cm). **Where found:** Fields, roadsides. Throughout. **Flowers:** April–Oct.
WHITE CLOVER, *T. repens.* White or tinged with pink. Leaves and flowerhead on separate stalks from a creeping runner. Leaflets with *pale triangular markings.* 4–10 in. (10–25 cm). **Where found:** Roadsides, lawns. Throughout. **Flowers:** April–Oct.

WHITE SWEET CLOVER, MELILOT   Young leaves,
Melilotus alba   seeds
A tall, smooth plant with long tapering spikes of tiny pealike flowers. Leaflets narrow; in 3's; with a sweet vanillalike odor when crushed. 2–8 ft. (0.6–2.4 m). See also Yellow Sweet Clover, p. 80. **Where found:** Roadsides, field edges, waste ground. Throughout. **Flowers:** April–Oct.
Use: Salad, cooked green, flavoring. The *young leaves,* before the flowers appear, can be added to salads or boiled for 5 min. The pealike *seeds* can be used to flavor soups and stews. The *dried leaves* can be used as a vanillalike flavoring for pastries. Rich in protein.   SPRING (leaves); SUMMER (seeds)

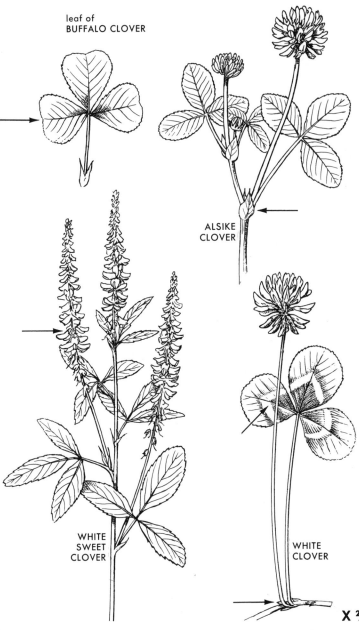

leaf of
BUFFALO CLOVER

ALSIKE
CLOVER

WHITE
SWEET
CLOVER

WHITE
CLOVER

X ⅔

MISCELLANEOUS COMPOSITE FLOWERS

 PILEWORT, FIREWEED Young leaves
Erechtites hieracifolia
A leafy, rank-smelling weed topped by a drab cluster of brush-shaped rayless flowers. The white hairlike disk flowers barely emerge from the *swollen-based* cylinder of green bracts. Stem grooved, unbranched. Leaves 3–8 in. (7.5–20 cm), thin, jagged-edged, alternate; highly variable. 1–8 ft. (0.3–2.4 m). See also Fireweed, *Epilobium,* p. 98. **Where found:** Moist soil, waste places, burns. Most of our area. **Flowers:** July–Sept.
Use: Salad, cooked green. In Asia the young leaves are eaten either raw or cooked. The strong flavor suggests that this is an acquired taste. Summer

GALINSOGA Leaves
Galinsoga spp.
Inconspicuous low weeds with slender, forking stems. Flower-heads ¼ in. (6 mm) across, with 5 tiny 3-lobed rays and a golden central disk. Leaves opposite, broad, coarsely-toothed; lower leaves stalked. Species vary from coarsely hairy to nearly hairless. 6–18 in. (15–45 cm). **Where found:** Cultivated ground, waste places. Widespread. **Flowers:** May–Nov.
Use: Cooked green. Excellent boiled for 10–15 min. and served with butter or vinegar. Summer

OX-EYE DAISY Young leaves
Chrysanthemum leucanthemum
The familiar white daisy of field and wayside. Biennial; producing a low rosette of leaves the first year, flowerstalks the second. Flowers 2 in. (5 cm) across; central disk yellow, depressed in middle. Leaves smooth, narrow, dark green, irregularly lobed. 1–3 ft. (30–90 cm). **Where found:** Fields, waste places. Throughout. **Flowers:** April–Aug.
Use: Salad. The tender young leaves (lighter green) make an interesting addition to salads. Spring

WILD CHAMOMILE Flowers
Matricaria chamomilla
A low, slender-stemmed plant with small daisylike flowers and sparse, feathery leaves. Leaves and stems apple-green, *pineapple-scented.* 6–12 in. (15–30 cm). **Where found:** Road edges, waste places. E. Canada, n. U.S.; locally abundant. **Flowers:** May–Oct.
Use: Tea. The dried flowers make a delicate, pale golden tea. Summer

CHICORY Young leaves, roots
Cichorium intybus **Color pl. 7**
Usually blue, but sometimes white or pink. See p. 144.
Early Spring (leaves); Fall–Early Spring (roots)

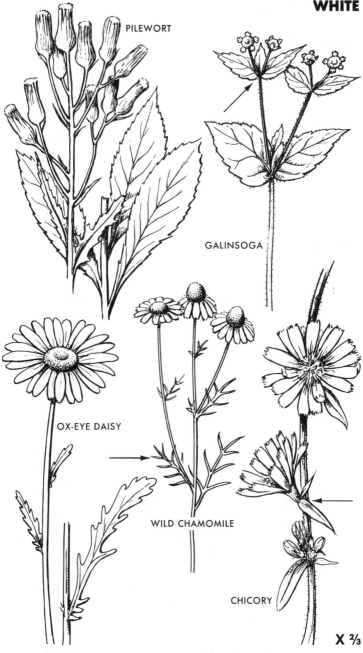

PILEWORT

GALINSOGA

OX-EYE DAISY

WILD CHAMOMILE

CHICORY

X ⅔

SHOWY CUPLIKE FLOWERS

  AMERICAN LOTUS, NELUMBO Young leaves, seeds,
Nelumbo lutea **tubers**
Flowers pale yellow, large; leaves bowl-shaped, huge — 1-2 ft.
 (30-60 cm) — usually held a few feet above the water. Seedpod
showerheadlike; each hole in its top contains a large seed.
Where found: Quiet water. Texas to Fla., north locally Minn.
 to Mass.; mostly west of Appalachians. **Flowers:** July–Sept.
Use: Potato, cooked green, cooked vegetable, nuts, flour. Although difficult to collect, the *tuberous enlargements* of the
 rootstock are excellent baked like sweet potatoes. The unrolling *young leaves* can be prepared like spinach. The immature
seeds can be eaten whole, either raw or cooked. The kernels
 from the ripe seeds (pierce the tough jacket, roast, and winnow)
can be eaten like nuts, or ground into flour. SPRING (leaves);
SUMMER–FALL (seeds); FALL–EARLY SPRING (tubers)

 PRICKLY-PEAR Young leaf pads, fruit,
Opuntia humifusa **seeds**
Note the *jointed pads* tufted with bristles and showy yellow
flowers. Fruit dull red, pulpy. Variable; usually prostrate.
Where found: Rocks, dry sandy soil. Minn. to Mass., south to
Okla., S.C. **Flowers:** May–July. **Fruit:** Aug.–Oct.
Use: Fruit, flour, thickener, cooked vegetable. The fleshy pulp
between the skin and seeds of the *fruit* is excellent chilled;
remove the bristles with a damp cloth. The dried *seeds* can be
ground to use as flour or a soup thickener. The tender *pads* can
be peeled, prepared like green beans. **Warning:** Handle with
gloves. SPRING–EARLY SUMMER (pads);
LATE SUMMER–FALL (fruit, seeds)

 YELLOW POND-LILIES Seeds, rootstock
Nuphar spp.
Familiar water-lilies with globelike yellow flowers, each flower
with a prominent disklike stigma in its center. Leaves with a
conspicuous notch between 2 rounded basal lobes.
Use: Potato, popcorn, flour, corn. The large *rootstocks* can be
used like potatoes; if the flavor is too strong, boil in 2 or 3
changes of water. Removed from their spindle-shaped pods, the
large *seeds* can be fried like popcorn, or parched and the win-
nowed kernels ground into flour or creamed like corn.
LATE SUMMER–FALL (seeds); FALL–EARLY SPRING (rootstock)
BULLHEAD-LILY, *N. variegatum.* **Color pl. 3.** Leaves
floating. **Where found:** Still water. Canada south to Iowa,
Md. **Flowers:** May–Sept.
SPATTERDOCK, COW-LILY, *N. advena* (not shown).
Leaves *erect,* rarely floating. **Where found:** Quiet water.
S. U.S. north to Wisc., s. New England. **Flowers:** May–Oct.

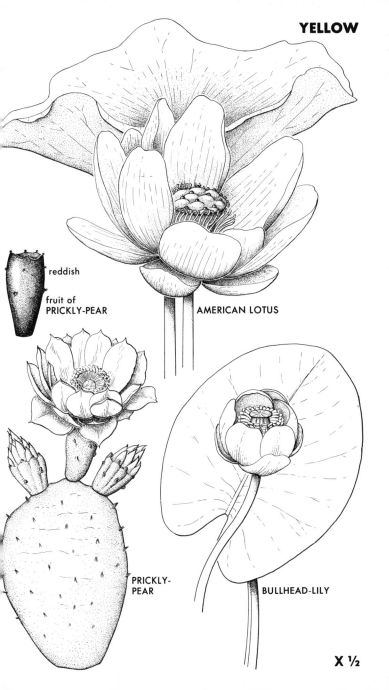

YELLOW

reddish

fruit of
PRICKLY-PEAR

AMERICAN LOTUS

PRICKLY-
PEAR

BULLHEAD-LILY

X ½

CLUBLIKE SPADIX WITH MINUTE FLOWERS; PARALLEL-VEINED LEAVES; WET PLACES

GOLDEN CLUB   **Dried seeds, dried rootstock**
Orontium aquaticum
Tiny bright yellow flowers cover a fingerlike *golden spadix* at the end of a naked fleshy stalk that juts conspicuously above the water's surface. Flowers followed by beanlike green seeds. The spathe is very small and soon disappears. The long-stalked elliptical leaves often float on the water. 1–2 ft. (30–60 cm). **Where found:** Swamp water, ponds, shores. Ky., W. Va., s. New England, south to La. and Fla. (commonest on or near the coastal plain). **Flowers:** March–June.
Use: Flour, cooked vegetable (peas). Once they have been thinly sliced and thoroughly dried, the stout starchy *rootstocks* can be ground into a nutritious, meal-like flour; the rootstocks are often deeply buried and difficult to dig. Once they have been thoroughly dried, the blue-green *seeds* can be ground into flour or boiled for 45 min. in several changes of water and served like peas. **Warning:** Both the rootstocks and seeds have the same acrid, biting quality as Jack-in-the-pulpit (p. 156) and cannot be eaten raw. Only thorough drying, not boiling, removes the acrid pepperiness.   LATE SUMMER–FALL (seeds);
FALL–EARLY SPRING (rootstock)

SWEETFLAG, CALAMUS   **Rootstock, young shoots**
Acorus calamus   **Color pl. 6**
Note the *yellow-green* irislike leaves and spicy smell of the leaves and rootstock. A tapering, fingerlike, yellow-green spadix juts at an angle from the side of a 2-edged stem that resembles the leaves. Usually found in large colonies. Often an entire colony will be without flowers. 1–4 ft. (0.3–1.2 m). **Where found:** Wet places, borders of quiet water. Most of our area. **Flowers:** May–Aug.
Use: Candy, salad. The stout, horizontal *rootstocks* can be candied. Peel, cut into ½ in. (1.3 cm) lengths, and boil in 4 or 5 changes of water until tender — about 1 hr. Then simmer for 20 min. in a rich sugar syrup and set out to dry. The tender inner parts of the spring *shoots* — up to 12 in. (30 cm) high — make a spicy addition to salads. **Caution:** Do not confuse with Blue Flag (p. 130) which is quite poisonous. Although superficially similar and growing in the same habitat, the leaves of Sweetflag are glossy, yellowish green, and aromatic, while those of Blue Flag are dull, blue-green, and odorless.   SPRING

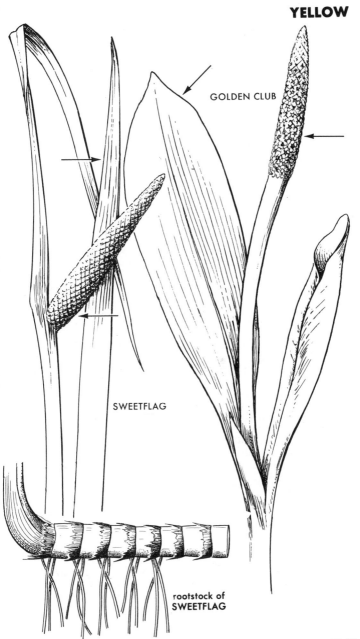

YELLOW

GOLDEN CLUB

SWEETFLAG

rootstock of
SWEETFLAG

X ⅔

4 PETALS; TERMINAL CLUSTERS: MUSTARDS

WINTER CRESS   **Young leaves, flowerbuds**
Barbarea vulgaris   **Color pl. 8**
Similar to the wild mustards (below) but leaves dark green, *glossy*. Basal leaves with rounded, earlike lobes; *upper leaves broad, toothed, clasping.* 1–2 ft. (30–60 cm). **Where found:** Moist waste ground. E. Canada south to Mo., Va. **Flowers:** April–Aug.
Use: Salad, cooked green, cooked vegetable. The *young leaves* form dense rosettes during warm spells in late winter; picked while the nights are still frosty, they are excellent added to salads or cooked like spinach. As the weather warms up and the leaves start becoming bitter, they should be boiled in 2 or 3 changes of water. The tight clusters of *flowerbuds* that appear after the leaves become too bitter to use can be boiled for 5 min. in 2 changes of water and served like broccoli.
LATE WINTER–EARLY SPRING (leaves); SPRING (buds)

WILD MUSTARDS   **Young leaves, flowerbuds,**
Brassica spp.   **young seedpods, seeds**
Note the conspicuous terminal clusters of 4-petaled yellow flowers. Lower leaves broad, deeply lobed; seedpods slender, ascending, ending in a conspicuous beak. A large, widespread group of edible plants; 2 representative species shown.
Use: Salad, cooked green, cooked vegetable, pickle, seasoning. The tender young basal *leaves,* gathered during the first warm days of spring, are excellent boiled for 30 min.; like spinach, they lose bulk upon cooking. Although slightly bitter, a few finely-chopped young leaves are a fine addition to fresh salads. As the leaves mature, they become too bitter to use. The clusters of unopened *flowerbuds,* boiled for 3–5 min., are a delicious substitute for broccoli; do not overcook. When gathering the flowerbuds, be sure to avoid including any of the extremely bitter upper-stem leaves. The tender *green seedpods,* collected while flowers are still in bloom, can be pickled or added fresh to salads. The ripe *seeds* are a familiar seasoning in pickle recipes; in addition, they make an outstanding hot yellow mustard when finely ground. Entire plant rich in vitamins A, B_1, B_2, and C; flowerbuds also rich in protein.
EARLY SPRING (leaves); SPRING–EARLY SUMMER (buds); SUMMER (tender pods, seeds)
FIELD MUSTARD, RAPE, *B. rapa.* Smooth, succulent, gray-green, dusted with whitish powder. Stem leaves typically with 2 roundish lobes that *clasp* the stem. **Where found:** Waste ground. Throughout. **Flowers:** April–Oct.
BLACK MUSTARD, *B. nigra.* Lower leaves irregularly lobed, bristly; upper leaves narrow, hairless. Seedpods held close to stem. 2–3 ft. (60–90 cm). **Where found:** Waste places, fields. Throughout. **Flowers:** April–Nov.

64

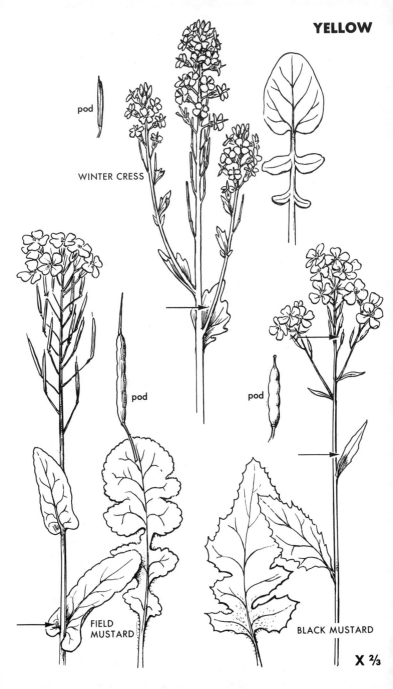

YELLOW

pod

WINTER CRESS

pod

pod

FIELD MUSTARD

BLACK MUSTARD

X ⅔

4 OR 5 PETALS; MISCELLANEOUS

 COMMON EVENING-PRIMROSE   **Young leaves, roots**
Oenothera biennis   **Color pl. 7**
A leafy, rough-hairy, biennial plant which produces a low
rosette of leaves the first year and a flowerstalk the second;
flowerstalks often reddish-stemmed, branched. Note the conspicuous yellow flowers at the end of a slender calyx tube rising
from a swollen ovary. Flowers 4-petaled, with an *x-shaped*
stigma and *reflexed* sepals. Flowers open toward evening and
wilt the next day. 1–5 ft. (0.3–1.5 m). **Where found:** Dry soil,
roadsides, waste ground. Canada south to Texas, Ark., Tenn.,
and n. Fla. **Flowers:** June–Oct.
Use: Cooked vegetable, salad, cooked green. Peel and boil the
pale pink first-year *taproots* for 20–30 min. in 2 or 3 changes of
water and serve with butter; mildly pungent, excellent. Locate
the first-year rosettes by looking near collections of the new or
old flowerstalks. The fresh roots are usually quite peppery, but
become milder during certain times of the year; generally these
times are late fall and early spring, but they can vary depending on the locality. The tender new *leaves* can be peeled and
served as a peppery addition to salads, or boiled for 20 min. in 2
or 3 changes of water and served with butter; slightly bitter but
palatable.   FALL (roots)–EARLY SPRING (roots, leaves)

YELLOW BEDSTRAW   **Leaves and flowers**
Galium verum
Note the whorls of 6–8 narrow leaves and dense clusters of tiny,
fragrant, 4-lobed, yellow flowers. 8–30 in. (20–75 cm). **Where
found:** Dry fields, roadsides. S. Canada south to Kans., Mo.,
Ind., Ohio, W. Va., and Va. **Flowers:** June–Aug.
Use: Rennet. An extract of the leaves and flowers was at one
time used to curdle milk for making cheese.   SUMMER

 WILD PARSNIP   **Roots**
Pastinaca sativa
Biennial, producing a low rosette of leaves in the first year and
a tall, stout, deeply grooved stalk topped by umbrellalike clusters of tiny 5-petaled golden flowers in the second. Leaves
divided into 5–15 stalkless, sharply toothed, ovate leaflets. 2–5
ft. (0.6–1.5 m). **Where found:** Waste places, roadsides.
Throughout. **Flowers:** May–Oct.
Use: Raw or cooked vegetable. The fleshy taproots from the
first-year plants can be excellent eaten raw, sautéed in butter,
or boiled until tender (as long as 30–45 min.). **Warning:** The
combination of wet or sweaty skin, contact with the leaves, and
exposure to sunlight may cause phytophotodermatitis. The
symptoms are like those of Poison Ivy, but the affected area
may remain reddened for months.   FALL–EARLY SPRING

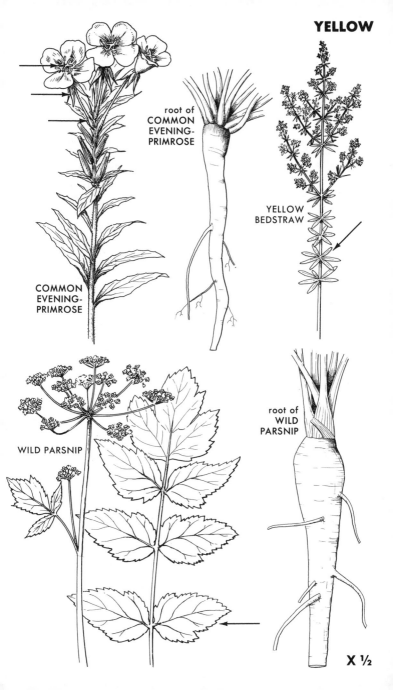

YELLOW

root of
COMMON
EVENING-
PRIMROSE

YELLOW
BEDSTRAW

COMMON
EVENING-
PRIMROSE

WILD PARSNIP

root of
WILD
PARSNIP

X ½

5-LOBED, BELL-LIKE FLOWERS

COMFREY   **Young leaves**
Symphytum officinale
A coarse, hairy perennial. Leaves alternate, spear-shaped, tapering into a *winged* stem. Flowers 5-lobed, bell-like, yellow, whitish, pinkish, or dull purple; in a 1-sided, curling cluster with a *pair of winglike leaves.* 2–3 ft. (60–90 cm). **Where found:** Roadsides, ditches, waste places. W. Ontario to Newfoundland, south to La., Tenn., Ga. **Flowers:** June–Sept.
Use: Cooked green, tea. Prepare the tender young leaves like spinach; the hairiness is reduced upon cooking. Older leaves are somewhat bitter and must be boiled in several changes of water. Steep the dried leaves to make tea.   EARLY SPRING

GROUND-CHERRIES, HUSK-TOMATOES   **Mature**
Physalis spp.   **fruit (only)**
Coarse-leaved plants with nodding, shallow-lobed, bell-like flowers, often with a dark center, that hang singly from leaf axils or forks in the stem. Leaves alternate. Sweet yellow, reddish, or purplish berrylike fruit are *enclosed in a papery bladder* formed by the sepals. CHINESE LANTERN-PLANT, *P. alkekengi,* is cultivated. Numerous species; 4 shown.
Use: Fresh or cooked fruit, jam. The ripe berries taste pleasantly sweet. Delicious raw, or made into pie, jam, preserves, or syrup; pectin must be added when making jam. The husk, or "lantern," and its berry often drop before the berry is fully ripe, but it ripens while on the ground. **Warning:** Leaves and *unripe* fruit poisonous.   LATE SUMMER–FALL
SMOOTH GROUND-CHERRY, *P. subglabrata.* Smooth, with diamond-shaped leaves that taper into leafstalks. Leaves often with 1 or 2 blunt teeth. Flower yellow with purplish throat. Fruit reddish or purplish. 3–5 ft. (0.9–1.5 m). **Where found:** Fields, roadsides, waste places. Minn., Ontario, s. New England, south to Texas and Ga. **Flowers:** July–Sept.
VIRGINIA GROUND-CHERRY, *P. virginiana.* Vaguely toothed or toothless leaves taper at both ends. Stem with long soft hairs. Flower yellow with purplish spots. Fruiting bladder deeply dented at base. Fruit red. 18–36 in. (45–90 cm). **Where found:** Woods, clearings. S. Manitoba, s. Ontario, s. New England, south to Texas and n. Fla. **Flowers:** April–Aug.
CLAMMY GROUND-CHERRY, *P. heterophylla.* Stems sticky-hairy; leaves with rounded bases, few teeth. Variable. Flower greenish yellow, brown center. Fruit yellow. 1–3 ft. (30–90 cm). **Where found:** Dry woods, clearings. Saskatchewan to New England and south. **Flowers:** May–Aug.
STRAWBERRY-TOMATO, *P. pruinosa.* Extremely downy, gray-green. Leaves coarsely toothed, heart-shaped at base. Fruit yellow. **Where found:** Dry soil, fields. Wisc., Ohio, N.Y., sw. Maine, south to Mo. and nw. Fla. **Flowers:** July–Oct.

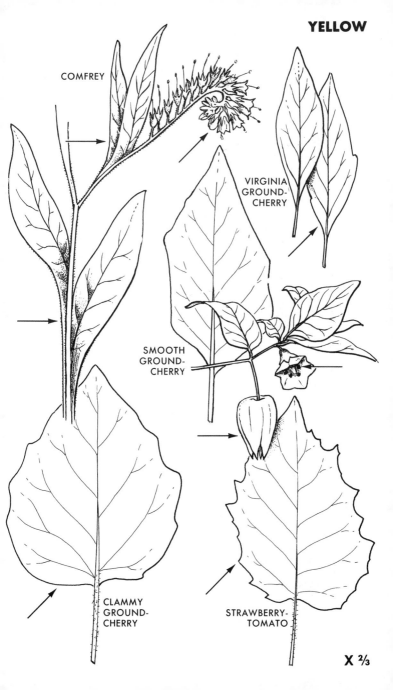

COMFREY

VIRGINIA GROUND-CHERRY

SMOOTH GROUND-CHERRY

CLAMMY GROUND-CHERRY

STRAWBERRY-TOMATO

X ⅔

5 (OR MORE) GLOSSY PETALS;
BUTTERCUPLIKE FLOWERS

 COMMON or TALL BUTTERCUP　　**Poisonous**
Ranunculus acris
Note the 5 (to 7) *glossy,* overlapping petals. Erect, branching, hairy. Basal leaves *deeply cut,* palm-shaped, with 5-7 unstalked segments. 2-3 ft. (60-90 cm). **Where found:** Fields, meadows. Minn. to s. Labrador, south to ne. Kans., Mo., Ill., Ind., Ohio, and Va. **Flowers:** May-Aug.
Warning: Buttercups contain varying amounts of an acrid poison that can cause intestinal irritation if eaten or skin blisters if handled.

SHRUBBY CINQUEFOIL　　**Leaves**
Potentilla fruticosa　　**Color pl. 6**
A bushy shrub that has 5- (to 7-) part leaves with toothless leaf segments. Leaves silky, whitish beneath. Stems *woody,* bark often loose. 1-3 ft. (30-90 cm). **Where found:** Meadows, shores. Across Canada and south to n. Iowa, Ill., Ohio, Penn., and n. N.J. **Flowers:** June-Oct.
Use: Tea. Steep the dried leaves for 5-10 min.　　SUMMER

MARSH-MARIGOLD, COWSLIP　　**Young leaves,**
Caltha palustris　　**Color pl. 2**　　**flowerbuds**
Flowers buttercuplike but larger — 1-1½ in. (2.5-3.8 cm) — with 5-9 shiny, deep yellow, petal-like sepals. Leaves deep green, glossy; *roundish* or broadly heart-shaped. Stem stout, hollow, succulent. 8-24 in. (20-60 cm). **Where found:** Swamps, brooks. Canada south to Neb., Iowa, Tenn., and S.C. **Flowers:** April-June.
Use: Cooked green, pickle. Collected before the plant has finished blossoming, the *young leaves* (with the stalks removed) are excellent cooked for 20-30 min. in 2 or 3 changes of boiling water. The *flowerbuds* can be used as capers when boiled for 10 min. in 2 changes of water and pickled in hot vinegar; *do not drink the juice* in which the buds have been pickled. **Warning:** Do not eat raw; Cowslip contains an acrid poison that is only dispelled upon cooking.　　EARLY SPRING-SPRING

 SILVERWEED　　**Roots**
Potentilla anserina　　**Color pl. 1**
Prostrate, spreading by slender runners. Flowers and leaves on separate stalks; leaves *silvery beneath,* with 7-25 paired, serrated leaflets that increase in size toward the tip. Colonial. 1-3 ft. (30-90 cm). **Where found:** Wet sandy shores, saltmarshes. Canada south to Iowa, Great Lakes, coastal New England. **Flowers:** June-Aug.
Use: Cooked vegetable. The fleshy roots can be roasted, added to soups and stews, or boiled for 20 min. and served with butter. Delicious; somewhat similar to parsnips or sweet potatoes in taste.　　FALL-EARLY SPRING

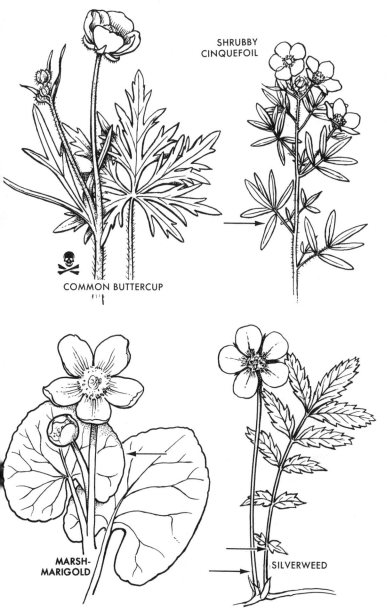

SHRUBBY CINQUEFOIL

COMMON BUTTERCUP

MARSH-MARIGOLD

SILVERWEED

X ⅔

5 PETALS; MISCELLANEOUS

YELLOW WOOD-SORRELS   **Tender leaves**
Oxalis spp.
Note the *sour taste* of the pale green, cloverlike leaflets. Flow-
ers pale yellow. Several species, all edible; 1 shown.
Use: Salad, cold drink. See wood-sorrels, p. 104.

SPRING–FALL

YELLOW WOOD-SORREL, *O. stricta.* Note the *sharp angle*
formed by the erect seedpods and their bent stalks. 6–15 in.
(15–37.5 cm). **Where found:** Waste ground. Widespread.
Flowers: May–Oct.

PURSLANE   **Stems and leaves, seeds**
Portulaca oleracea
Prostrate, with succulent light green leaves and reddish stems.
Note the rosettes of fleshy, paddle-shaped leaves, each with a
small, stalkless, 5-petaled flower that opens only on sunny
mornings. **Where found:** Rich, sandy soils; waste places.
Throughout. **Flowers:** May–Nov.
Use: Salad, cooked green, pickle, flour. The mildly acid, mucilaginous *stems and leaves* are excellent added fresh to salads or
boiled for 10 min. in just enough water to cover; wash them
thoroughly first to remove any sand or grit. The tender fat
stems can be pickled. The tiny black *seeds* can be made into a
nutritious flour. The seeds will continue to ripen even though
the plants are no longer in the ground; to gather them in
quantity, place the plants on a sheet, dry for a few weeks, then
pound and sift the dried plants through a strainer to separate
the seeds from the rest. Leaves and stems rich in iron; also
contain vitamins A and C, calcium, phosphorus.

SUMMER (stems, leaves); LATE SUMMER (seeds)

COMMON MULLEIN   **Leaves**
Verbascum thapsus   **Color pl. 10**
Biennial, producing a low rosette of leaves in the first year and
a stout stalk topped by a clublike flowerhead in the second.
Leaves large, densely woolly, gray-green; upper leaves taper
into stem. 2–6 ft. (0.6–1.8 m). **Where found:** Poor fields, waste
places. Most of our area. **Flowers:** May–Sept.
Use: Tea. Steep the dried leaves for 5–10 min.   SUMMER

WHORLED LOOSESTRIFE   **Leaves**
Lysimachia quadrifolia
Delicate, erect; leaves and flowers *in whorls of 4 (3–6)*. Flowers
on slender stalks in leaf axils and dotted with red near center.
1–3 ft. (30–90 cm). **Where found:** Open woods, sunny thickets,
shores. Wisc., s. Ontario, Maine, south to Ill.; in uplands to
Ala., Tenn., and Ga. **Flowers:** June–Aug.
Use: Tea. Prepare the dried leaves like oriental tea. Flat
tasting; mix with other teas.   SUMMER

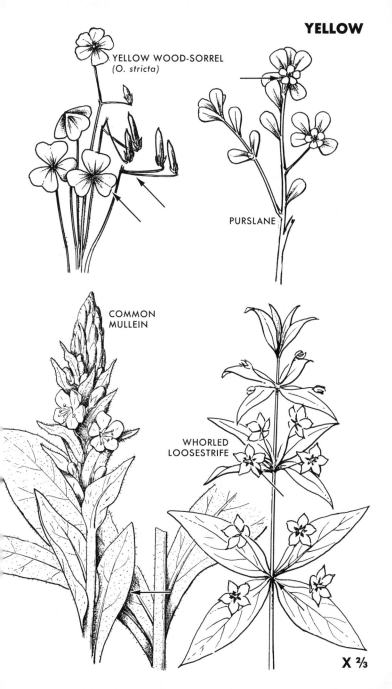

YELLOW WOOD-SORREL
(O. stricta)

PURSLANE

COMMON
MULLEIN

WHORLED
LOOSESTRIFE

X ⅔

6 PETALS: LILY FAMILY (1)

TROUT-LILY, DOG-TOOTH VIOLET   **Young leaves,**
Erythronium americanum   **Color pl. 14**   **corm**
Petals reflexed; yellow, often brown-purple beneath. Basal
leaves (2) smooth, *mottled*. Colonial. 4–10 in. (10–25 cm).
Where found: Rich moist woods, alluvial bottoms. Minn. to
Nova Scotia, south to Okla., Ga. **Flowers:** March–May.
Use: Cooked green, cooked vegetable. The very young *leaves*
can be boiled for 10–15 min. and served with vinegar. The
deeply buried, bulblike *corms* can be boiled for 20–25 min. and
served with butter. One of our more attractive spring flowers;
use the bulbs sparingly and only when abundant. **Warning:**
Trout-lily may be mildly emetic.   Early Spring

CLINTONIA, CORN-LILY   **Young leaves**
Clintonia borealis   **Color pl. 15**
2 or 3 shining leaves clasp the base of a leafless stalk topped by
several yellow-green nodding bells. Berries deep blue, inedible.
Young leaves *taste of cucumber*. 6–16 in. (15–40 cm). **Where
found:** Cool woods; open slopes on mts. Canada, n. U.S.; south
in mts. to Tenn. and Ga. **Flowers:** May–Aug.
Use: Salad, cooked green. Before fully unfurled, the young
leaves can be added to salads or boiled for 10 min. and served
with butter. As the leaves mature, the cucumber taste becomes
harsh and unpleasant.   Early Spring

CANADA or WILD YELLOW LILY   **Bulb**
Lilium canadense   **Color pl. 6**
Note the spotted, lemon-yellow to orange or red, nodding bells
and whorls of narrow, parallel-veined leaves. Bulbs waxy, with
numerous scalelike bulblets. 2–5 ft. (0.6–1.5 m). **Where
found:** Rich moist soil, wet meadows, ditches. Se. Canada,
ne. U.S.; in uplands to Ky., Va. **Flowers:** June–Aug.
Use: Cooked vegetable. The fleshy bulbs can be roasted, added
as an okralike thickener to soups and stews, or boiled for 20
min. and served with butter. **Note:** Although the bulbs of most
true lilies, *Lilium* spp., are edible, few are found in any abundance; this species and Turk's-cap Lily, p. 92, are occasional
exceptions.   Summer

INDIAN CUCUMBER-ROOT   **Tuber**
Medeola virginiana
Note the slender stem with *2 whorls* of leaves. Flowers greenish
yellow, dangling, with reflexed tips and reddish stamens. Tuber
crisp, waxy looking, cucumber-flavored. Berries blue, inedible.
1–3 ft. (30–90 cm). **Where found:** Moist woods. Minn. to
Nova Scotia, south to La., Fla. **Flowers:** May–June.
Use: Salad, pickle. Outstanding. **Note:** Gather only when
found in abundance.   Spring–Fall

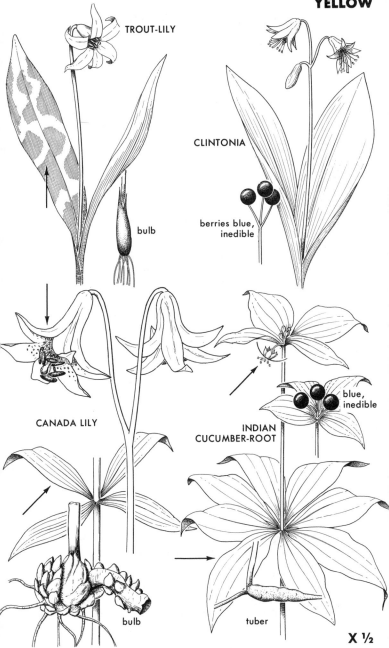

YELLOW

TROUT-LILY

bulb

CLINTONIA

berries blue,
inedible

CANADA LILY

bulb

INDIAN
CUCUMBER-ROOT

blue,
inedible

tuber

X ½

6 PETALS; LILY FAMILY (2)

 ASPARAGUS **Young shoots**
Asparagus officinalis
A familiar garden escape. Soft, feathery bunches of threadlike green branchlets cluster in the axils of brownish scalelike leaves. Side branches evenly spaced along a central stalk; taper upwards in a Christmas-tree shape. Flowers tiny, greenish yellow, dangling on slender stems along side branches. Berries red. 2–6 ft. (0.6–1.8 m). **Where found:** Roadsides, fields, fencerows. Most of our area. **Flowers:** May–June.
Use: Asparagus. Steam or boil for 10–15 min. Excellent. To find the new shoots look for last year's dead stalks.
EARLY SPRING

 BELLWORTS **Young shoots**
Uvularia spp.
Bell-like yellow or creamy flowers droop at the end of a forking leafy stem. Leaves parallel-veined, stalkless. Some species with leaves that clasp or surround the stems.
Use: Asparagus. As with solomon's-seals (below), discard the leafy portions of the young shoots and boil for 10 min. **Note:** Gather only when abundant. EARLY SPRING
WILD OATS, SESSILE BELLWORT, *U. sessilifolia*. **Color pl. 14.** Flowers creamy. Leaves stalkless but do not surround stem; stem angled. 6–13 in. (15–32.5 cm). **Where found:** Moist woods. S. Canada south to Mo., Ga. **Flowers:** April–June.

 SOLOMON'S-SEALS **Young shoots, rootstock**
Polygonatum spp.
Clusters of greenish yellow bells dangle beneath parallel-veined leaves arranged alternately along an arching stem. Berries blue-black, inedible. Rootstock stout, whitish, with large circular "seals." See also False Solomon's-seal, p. 52.
 Use: Asparagus, salad, potato. The *young shoots* (minus the leafy heads, which turn bitter when cooked) can be boiled for 10 min. and served like asparagus. The whole shoots can be chopped up and added to salads. The starchy *rootstocks* can be added to stews, or boiled for 20 min. and served like potatoes. Gather only when abundant. **Caution:** Do not confuse the rootstocks with those of May-apple (p. 20) which also have large nodes but no circular "seals," or those of False Solomon's-seal (p. 52), which are slender and yellowish.
EARLY SPRING (shoots); ALL YEAR (rootstock)
SOLOMON'S-SEAL, *P. biflorum*. Flowers and berries in *pairs*. 1–3 ft. (30–90 cm). **Where found:** Woods, thickets. Great Lakes to Conn. and south. **Flowers:** April–June.
GREAT SOLOMON'S-SEAL, *P. canaliculatum* (not shown). Taller, coarser; flowers in larger clusters. Usually over 4 ft. (1.2 m). **Where found:** Woods, openings. S. Manitoba to e. New England, south to Okla., S.C. **Flowers:** May–June.

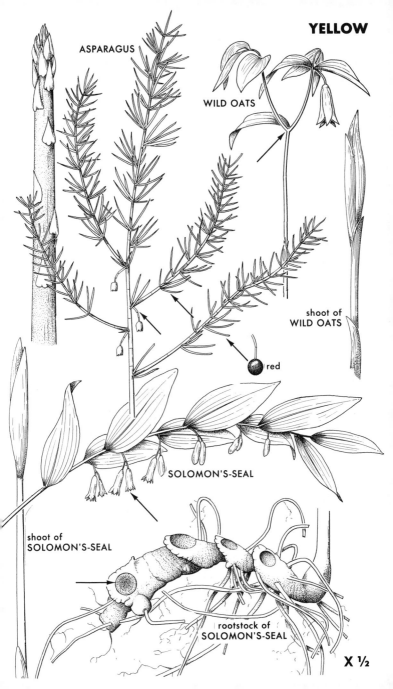

YELLOW

ASPARAGUS

WILD OATS

shoot of
WILD OATS

red

SOLOMON'S-SEAL

shoot of
SOLOMON'S-SEAL

rootstock of
SOLOMON'S-SEAL

X ½

MISCELLANEOUS LIPPED FLOWERS

NORTHERN FLY-HONEYSUCKLE   **Fruit**
Lonicera villosa   **Color pl. 4**
A small, erect honeysuckle with *downy-hairy* leaves. Leaves oval, in pairs; flowers paired on short stalks, tubular, lipped. Berries *blue*. 1–3 ft. (30–90 cm). **Where found:** Rocky or peaty soil, bogs, swamps. Manitoba to Newfoundland, south to northernmost U.S. **Flowers:** May–Aug. **Fruit:** June–Aug. **Use:** Fresh or cooked fruit, jelly. A little known but excellent fruit; prepare like blueberries.   SUMMER (June–July)

PALE TOUCH-ME-NOT, JEWELWEED
Impatiens pallida   **Young shoots;**
                          stems and leaves (external use)
The pendent blossoms are similar to those of Spotted Touch-me-not (p. 92) but *pale yellow*. Stems succulent, watery; leaves look *silvery when under water;* ripe seedpods spring open when touched lightly. 3–5 ft. (0.9–1.5 m). **Where found:** Wet shady places, calcareous mt. woods. Saskatchewan, Newfoundland south to Kans., Mo., Ga. (in mts.). **Flowers:** July–Oct. **Use:** Cooked green; remedy for Poison Ivy and nettles. The *young shoots* — up to 6 in. (15 cm) high — can be boiled for 10–15 min. in 2 changes of water and served as a cooked green; do not drink the cooking water. Washing with the raw juice from the crushed *stems and leaves* soothes the sting of nettles (p. 150) and is reputed to prevent the rash from Poison Ivy (p. 182).   EARLY SPRING (shoots); SUMMER (stems, leaves)

HORSEMINT   **Leaves**
Monarda punctata
Note the whorls of *gaping,* yellowish, purple-flecked flowers in the upper leaf axils, and the *distinctive white or pale purple bracts* at their base. Aromatic. 1–3 ft. (30–90 cm). **Where found:** Dry sandy soil, coastal plain, prairies. Minn. to Long Island and south. **Flowers:** June–Oct. **Use:** Tea. Prepare like oriental tea.   SUMMER

PUTTYROOT   **Corms**
Aplectrum hyemale
A single, broad, parallel-veined leaf — 4–7 in. (10–17.5 cm) — is produced in summer, overwinters, but disappears as the flowers develop. Flowers variously tinged with yellow, green, or purple. Lip crinkly-edged, whitish. Roots often a string of globular corms from which new leaves and flowerstalks appear. 10–16 in. (25–40 cm). **Where found:** Rich deciduous woods, moist soil. S. Canada south locally to Ark., Tenn., and Ga. **Flowers:** May–June. **Use:** Cooked vegetable. Boil the bulblike corms for 20 min. and serve with butter. **Note:** Relatively uncommon; use only in emergencies.   SPRING

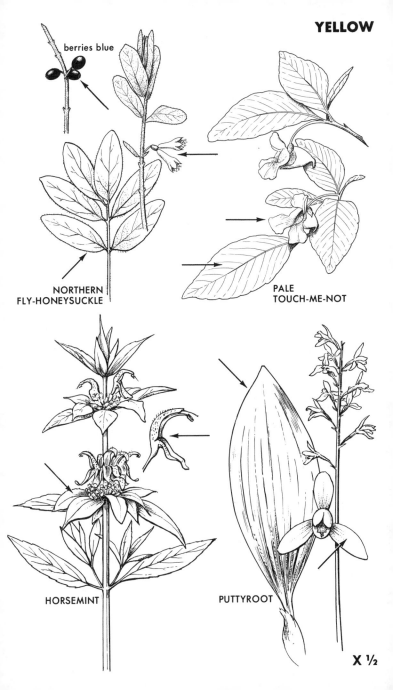

YELLOW

berries blue

NORTHERN
FLY-HONEYSUCKLE

PALE
TOUCH-ME-NOT

HORSEMINT

PUTTYROOT

X ½

PEALIKE FLOWERS; 3 LEAFLETS

Note: See also Scotch Broom, p. 82.

 BLACK MEDICK Seeds
Medicago lupulina
Prostrate; resembles Hop Clover (below), but note the longer leafstalks, downy stems, and tightly *coiled, one-seeded, black* pods. Leaflets often finely bristle-tipped. **Where found:** Roadsides, waste places. Throughout. **Flowers:** March–Dec. **Use:** Flour. The seeds can be parched and eaten out of hand or ground into flour. SUMMER-FALL

 WILD INDIGO Poisonous
Baptisia tinctoria
Much-branched, with smooth gray-green to bluish leaves and stems. Leaves nearly stalkless; turn *black* when dried. Flowers loosely clustered along ends of branches. 1–3 ft. (30–90 cm). **Where found:** Dry woods, clearings. Minn., s. Ontario, s. Maine, south to La. and Fla. **Flowers:** April–Sept. **Warning:** Although the tender young shoots somewhat resemble Asparagus and have been used as such in certain areas of New England, they have been known to poison cattle and are best left alone.

YELLOW SWEET CLOVER Young leaves, seeds
Melilotus officinalis
A tall, loosely branched plant with elongated spikes of tiny flowers. Leaves and stems smooth; leaflets narrow, with a delicate, vanillalike odor when crushed. Closely resembles White Sweet Clover, p. 56. 2–5 ft. (0.6–1.5 m). **Where found:** Roadsides, waste ground. Throughout. **Flowers:** April–Oct. **Use:** Salad, cooked green, flavoring. The *young leaves* (before the flowers appear) can be added to salads or boiled for 5 min. The pealike *seeds* can be used to flavor soups and stews. The crushed *dried leaves* can be used as a vanillalike flavoring for pastries. A good source of protein. SPRING (young leaves); SUMMER (seeds)

 HOP CLOVER Young leaves,
Trifolium agrarium flowerheads, seeds
Resembles Black Medick (above) but erect. Flowerheads ½–¾ in. (1.3–1.9 cm); several similar species in our area are smaller. After blooming, flowers droop, forming brownish heads that suggest dried hops. 6–16 in. (15–40 cm). **Where found:** Roadsides, waste places. Throughout. **Flowers:** May–Sept. **Use:** Salad, cooked green, flour, tea. See clovers, *Trifolium* spp., p. 56. SPRING-SUMMER

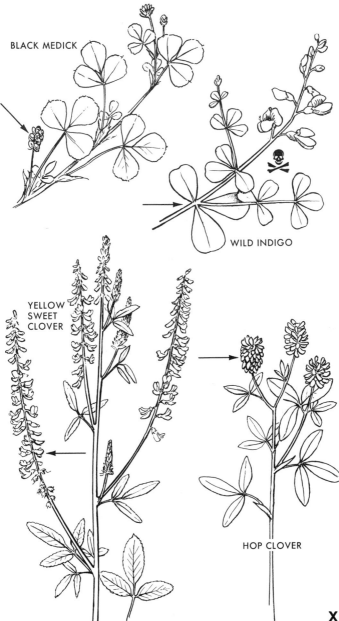

BLACK MEDICK

WILD INDIGO

YELLOW
SWEET
CLOVER

HOP CLOVER

X ⅔

PEALIKE FLOWERS;
MISCELLANEOUS LEAVES

SICKLEPOD   **Young shoots**
Cassia tora
Loosely constructed flowers are borne singly in the upper leaf
axils. Flowers 5-petaled, with unequal stamens and dark brown
anthers. Leaflets in 2 or 3 pairs. Seedpod *long, sickle-shaped.*
1½-2 ft. (45-60 cm). **Where found:** Rich soil, riverbanks,
waste ground. E. Kans., Mo., Ill., Mich., Ind., and Penn., south
to Texas and Fla. **Flowers:** July–Sept.
Use: Asparagus. The young shoots can be boiled for 10–15
min. in 2 or 3 changes of water and served with butter or
vinegar. The unpleasant odor of the fresh shoots disappears
when they are cooked.   SUMMER

SCOTCH BROOM   **Flowerbuds, young pods,**
Cytisus scoparius   **seeds**
A dense, *stiff-branched, deciduous or evergreen* shrub. Stems
angled, dark green; leaves small — up to ½ in. (1.3 cm) —
lance-shaped or 3-part, often dropping early. Flowers showy, 1
in. (2.5 cm); fruit flattened pods. 3-5 ft. (0.9-1.5 m). **Where
found:** Dry sandy soil, roadsides, thickets. W. New York,
sw. Maine, Nova Scotia, south to W. Va. and Ga., especially
near the coast. **Flowers:** May–June. **Fruit:** July–Sept.
Use: Pickle, coffee. See p. 182.   SUMMER

GOAT'S-RUE   **Poisonous**
Tephrosia virginiana
Note the silky whitish hairs. Flowers showy, *bicolored* (yellow
above, pink below). Leaves feather-compound with numerous
narrow leaflets. Seedpods long, flat, slender, hairy; seeds len-
til-like. 1-2 ft. (30-60 cm). **Where found:** Sandy woods,
openings. Minn., s. Ontario, s. N.H., south to Texas, Fla.
Flowers: May–Aug.
Warning: Do not use the seeds as a substitute for peas or
lentils. The root is highly poisonous, and the crushed stems
have been used as a fish poison.

RATTLEBOX   **Poisonous**
Crotalaria sagittalis
A small hairy plant. Leaves toothless, alternate, nearly stalkless and often with *arrowheadlike stipules* facing down the
stem. Seeds held in a black pod, rattle when dry. 12-16 in.
(30-40 cm). **Where found:** Dry sandy soils, fields, clearings.
Minn., Wisc., s. Mich., s. Ohio, s. N.Y., and Mass., south to
Texas and Fla. **Flowers:** June–Sept.
Warning: Some authors have suggested that the roasted seeds
can be used as a coffee substitute. However, the raw seeds are
poisonous, and while roasting or boiling may remove their
toxicity, it is best to leave them alone.

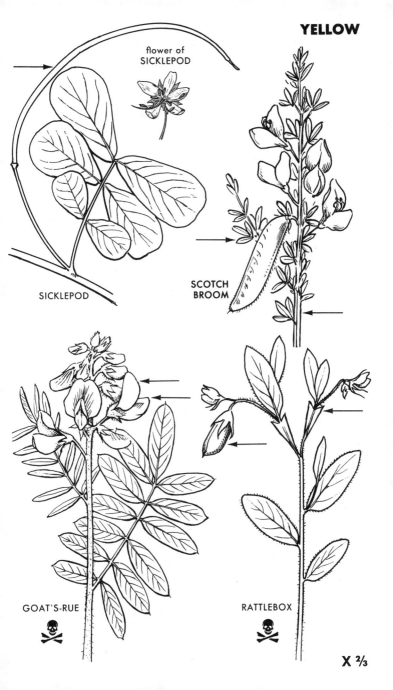

YELLOW

flower of
SICKLEPOD

SICKLEPOD

SCOTCH
BROOM

GOAT'S-RUE
☠

RATTLEBOX
☠

X ⅔

DANDELIONLIKE FLOWERS

COLTSFOOT   **Leaves**
Tussilago farfara
Flowers bristly, with numerous yellow rays in layers; stalks with reddish scales. The large basal leaves appear after the flowers. 6–18 in. (15–45 cm). **Where found:** Waste ground. South to Ohio, Penn., and N.J. **Flowers:** March–June.
Use: Candy (cough drops), cough syrup, tea, seasoning (salt). An excellent cough syrup or hard candy (cough drop) can be made by boiling the *fresh* leaves and adding sugar to the resultant extract. When making hard candy, add 2 cups of sugar for every cup of extract and boil until the rich syrup forms a hard ball when dropped in cold water. The *dried* leaves can be steeped to make a fragrant tea, or burned and the residue used as a saltlike seasoning.   Spring–Summer

COMMON DANDELION   **Young leaves, flower-**
Taraxacum officinale   **Color pl. 8**   **buds, flowers, roots**
The familiar lawn weed with solitary flowers and downy white seedballs. Leaves with sharp irregular lobes; stems milky, hollow. Outer bracts of flowers *reflexed.* 2–18 in. (5–45 cm). **Where found:** Lawns, roadsides. Throughout. **Flowers:** March–Sept.
Use: Salad, cooked green, cooked vegetable, fritters, coffee. The *young leaves,* gathered before the flowers appear, can be added to salads or boiled for 5–10 min. Although the entire leaf can be used, the blanched part just below soil level is best. Gathered when they are still tucked down in the rosette of leaves, the young *flowerbuds* can either be boiled for several minutes and served with butter, or pickled. The *flowers* are excellent dipped in batter and fried. To make a delicious coffeelike beverage, bake the *roots* in a slow oven until brown and brittle, grind, and perk like commercial coffee. Leaves rich in vitamin A.   Early Spring (leaves, buds);
Spring–Early Summer (flowers);
Fall–Early Spring (roots)

YELLOW GOAT'S-BEARD   **Young leaves, roots**
Tragopogon pratensis
The pale yellow blossom with its long slender bracts opens in the morning and closes by midday. Light green grasslike leaves clasp the smooth stem, which contains milky juice. Seedballs dandelionlike but much larger. 1–3 ft. (30–90 cm). See also Oyster-plant, p. 126. **Where found:** Waste places, roadsides, fields. S. Canada south to Ohio, N.J. **Flowers:** June–Oct.
Use: Salad, cooked green, cooked vegetable, coffee. The tender basal *leaves* can be eaten raw or cooked (boil 5 min.). The *roots* can be boiled for 20–30 min., or roasted and ground into coffee.
         Spring (leaves); Fall–Early Spring (roots)

84

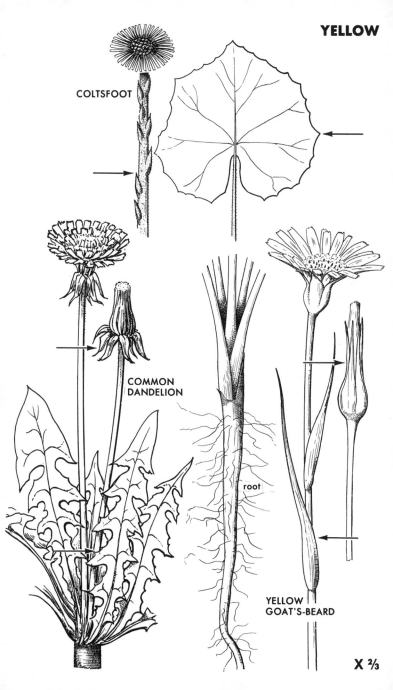

YELLOW

COLTSFOOT

COMMON DANDELION

root

YELLOW GOAT'S-BEARD

X ⅔

DANDELIONLIKE FLOWERS; TALL PLANTS

SOW-THISTLES   **Young leaves**
Sonchus spp.
Tall plants with *prickly-edged* leaves, bitter milky juice, and
dandelionlike blossoms in loose, somewhat flat-topped clusters.
Use: Salad, cooked green. The young leaves can be prepared
like those of Common Dandelion, p. 84. In the case of Spiny-
leaved Sow-thistle, remove the spines first.   SPRING
SPINY-LEAVED SOW-THISTLE, *S. asper.* Leaves *very
spiny;* curling at their bases and clasping a smooth, ridged
stem. 1-5 ft. (0.3-1.5 m). **Where found:** Waste places, roadsides. Nearly throughout. **Flowers:** April-Oct.
FIELD SOW-THISTLE, *S. arvensis.* Weak spines, *hairy*
bracts and stalks. 1½-4 ft. (0.4-1.2 m). **Where found:** Fields,
roadsides, waste places. Canada south to Mo., Ohio, and Md.
Flowers: July-Oct.
COMMON SOW-THISTLE, *S. oleraceus* (not shown). Hairless; leaves with sharp-pointed basal lobes that clasp the stem.
1-8 ft. (0.3-2.4 m). **Where found:** Waste places, fields, roadsides. Nearly throughout. **Flowers:** April-Oct.

WILD LETTUCES   **Young leaves,**
Lactuca spp.   **developing flowerheads**
Tall, leafy plants with numerous small, pale, dandelionlike
flowers in long loosely-branched clusters. Leaves dandelionlike,
scattered along stems. Sap milky, bitter. Numerous species; 3
shown. See also blue lettuces, p. 144.
Use: Salad, cooked green, cooked vegetable. Although somewhat bitter, the *young leaves* can be added fresh to salads or
boiled for 10-15 min. in 1 change of water and served with
butter or vinegar. The cooked leaves still leave a slightly bitter
aftertaste and are best mixed with other greens. The *developing flowerheads,* before the stems unfold and the flowers bloom,
impart a unique bitter flavor when added to casseroles.
       SPRING (leaves); SUMMER (developing flowerheads)
WILD LETTUCE, *L. canadensis.* Stems smooth, dusted with
white. Leaves extremely variable, deeply lobed to lance-
shaped. Lower leaves up to 10 in. (25 cm) long. 4-10 ft. (1.2-
3 m). **Where found:** Thickets, clearings in woods, roadsides.
S. Canada south to Texas, Mo., n. Fla. **Flowers:** June-Oct.
PRICKLY LETTUCE, *L. scariola.* Lower stem and leaves
prickly. Flower clusters sparser than in the above species. 2-7
ft. (0.6-2.1 m). **Where found:** Waste places, roadsides. Most of
our area. **Flowers:** June-Oct.
HAIRY LETTUCE, *L. hirsuta.* Resembles Wild Lettuce
(above) but less leafy. Flowers often reddish-yellow; stem
reddish. Lower stem and leaves often hairy. 1-6 ft.
(0.3-1.8 m). **Where found:** Dry open woods, clearings.
S. Canada south to Texas, La., and Va. **Flowers:** June-Oct.

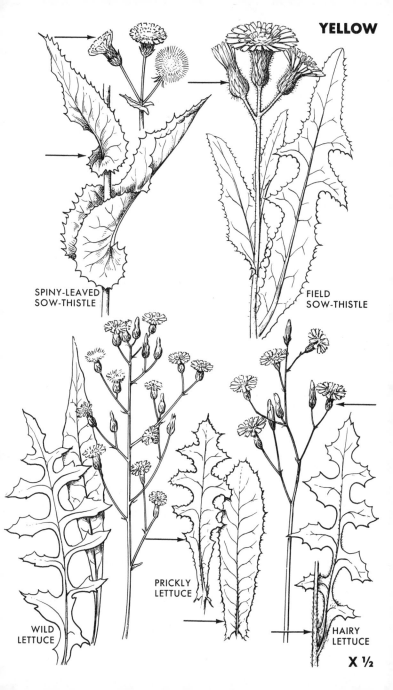

YELLOW

SPINY-LEAVED
SOW-THISTLE

FIELD
SOW-THISTLE

WILD
LETTUCE

PRICKLY
LETTUCE

HAIRY
LETTUCE

X ½

SUNFLOWERLIKE FLOWERS

COMPASS-PLANT, ROSINWEED Hardened sap
Silphium laciniatum
Tall; stem stout. Leaves *very large, deeply cut;* stiff, rough-bristly. 4–10 ft. (1.2–3 m). **Where found:** Prairies. N. Dak. to Mich., south to Okla., and Texas. **Flowers:** July–Sept.
Use: Chewing gum. Chew the hardened sap that forms after the flowerheads have been removed. **Note:** Several other large rosinweeds, *Silphium* spp., also produce edible sap.
SUMMER

COMMON SUNFLOWER Seeds
Helianthus annuus   **Color pl. 9**
Flowers *large* — up to 6 in. (15 cm) across — with a *brownish central disk* and golden rays. Leaves rough, slender-stalked, *heart- or spade-shaped;* alternate. Stems rough-hairy. 3–12 ft. (0.9–3.6 m). **Where found:** Waste places, fields, prairies. Minn. to Texas and west; local in East. **Flowers:** June–Oct.
Use: Nuts, cereal, flour, oil, coffee. The seeds are smaller than those from domesticated plants, but just as good. Boiling the crushed kernels releases a light vegetable oil. The roasted shells can be used to make coffee. LATE SUMMER–FALL

ELECAMPANE Rootstock
Inula helenium
Tall. Stems stout, hairy, mostly unbranched. Note the many narrow rays and broad bracts. Leaves woolly beneath; lower leaves stalked, upper leaves clasping. Rootstock spindle-shaped. 2–6 ft. (0.6–1.8 m). **Where found:** Roadsides, pastures. S. Ontario, Nova Scotia south. **Flowers:** July–Sept.
Use: Candy. Prepare like Wild Ginger (p. 96). Bitter-sweet, aromatic, addictive. FALL–EARLY SPRING

JERUSALEM ARTICHOKE Tubers
Helianthus tuberosus   **Color pl. 12**
A tall, coarse sunflower with broad, rough leaves and rough-hairy stems. Upper leaves alternate; lower leaves often opposite, occasionally in whorls of 3. Central disk of flowers yellow. Large, crisp tubers (often as large as a medium-sized potato) terminate slender runners that radiate from the base of each plant. Colonial. 6–10 ft. (1.8–3 m). **Where found:** Waste ground, damp thickets, fields. Saskatchewan, Ontario, south to Okla., Ark., Tenn., and Ga. **Flowers:** Aug.–Oct.
Use: Potato, salad, pickle. One of the best of the wild food plants. The fresh tubers can be substituted in recipes calling for potatoes; they can also be sliced into thin sections and added to salads, or boiled briefly and pickled for a few weeks in wine vinegar. Located while still in flower, Jerusalem Artichoke provides an easy abundance of tubers throughout the fall and winter whenever the ground is unfrozen.
FALL–EARLY SPRING

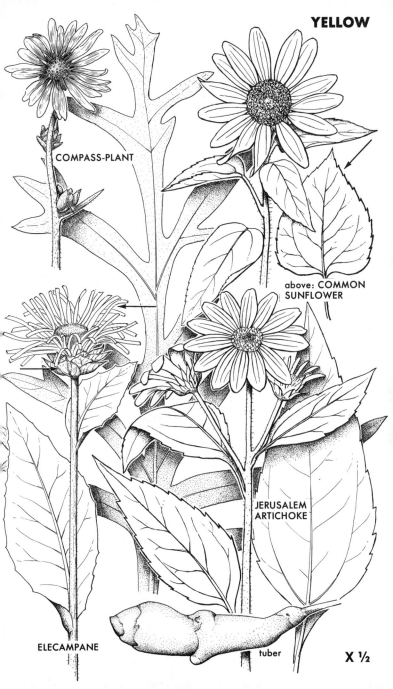

YELLOW

COMPASS-PLANT

above: COMMON
SUNFLOWER

ELECAMPANE

JERUSALEM
ARTICHOKE

tuber

X ½

MISCELLANEOUS COMPOSITE FLOWERS

PINEAPPLE-WEED   **Flowers**
Matricaria matricarioides
A low, inconspicuous weed somewhat resembling Wild Chamomile (p. 58), but leafier and flowers lacking the white rays. The finely-cut leaves and greenish-yellow flowerheads *smell of pineapple* when crushed. 6–18 in. (15–45 cm). **Where found:** Roadsides, waste places. S. Canada south to Mo., Ind., Ohio, Penn., and Del. **Flowers:** June–Oct.
Use: Tea. The fresh or dried flowers make a pale golden, pineapple-scented tea when steeped in hot water. Excellent.
                                                      SUMMER

SWEET GOLDENROD   **Leaves and flowers**
Solidago odora   **Color pl. 13**
Note the showy plumelike clusters of small yellow-rayed blossoms (each blossom with 3–5 rays) and the sweet, *aniselike odor* of the crushed leaves. The slender parallel-veined leaves are toothless, smooth, and show tiny transparent dots when held up to the sun. 1½–3 ft. (45–90 cm). **Where found:** Dry open woods, roadbanks, pine barrens. Mo., Ohio, N.Y., s. N.H. south to Texas and Fla. **Flowers:** July–Oct.
Use: Tea. The fresh or dried leaves and flowers brew up into a delicate, anise-flavored tea when steeped in hot water for 10 min. Excellent.   SUMMER

COMMON TANSY   **Young leaves, flowers**
Tanacetum vulgare
Note the small golden-yellow heads of disk flowers (ray flowers absent, or few and inconspicuous) arranged in showy flat-topped clusters. Leaves sharply-cut, fernlike, strongly aromatic. Stems *hairless*. 1–3 ft. (30–90 cm). **Where found:** Roadsides, waste ground. Much of our area. **Flowers:** June–Sept.
Use: Seasoning. The fresh young leaves and flowers can be used as a seasoning in lieu of sage. Use sparingly; it's quite strong. Best with fish. **Warning:** Although Tansy is quite safe if used in the tiny amounts needed for seasoning, it contains an oil that can be fatal when consumed in large quantities.
                SPRING (young leaves); SUMMER (flowers)

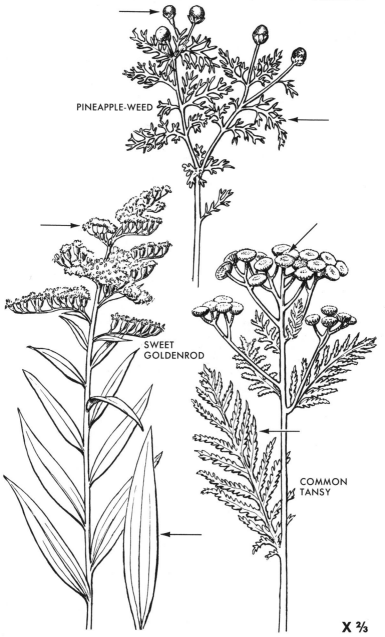

YELLOW

PINEAPPLE-WEED

SWEET
GOLDENROD

COMMON
TANSY

X ⅔

MISCELLANEOUS ORANGE FLOWERS

TURK'S-CAP LILY   **Bulb**
Lilium superbum
A showy orange-red lily with strongly reflexed spotted petals
(forming a "Turk's cap"), a *green central star,* and *protruding*
stamens. Leaves usually in whorls. Bulbs whitish, composed of
fleshy, scalelike bulblets. 3–8 ft. (0.9–2.4 m). **Where found:**
Meadows, moist or wet soil. N.Y. to New Brunswick, south to
Ala. and Ga. **Flowers:** July–Aug.
Use: Cooked vegetable. See Canada Lily, p. 74.   Summer

SPOTTED TOUCH-ME-NOT,   **Young shoots;**
JEWELWEED   **stems and leaves (externally)**
Impatiens capensis   **Color pl. 2**
Spotted orange blossoms dangle jewel-like at the end of slender
stalks. Stems succulent, watery. 2–5 ft. (0.6–1.5 m). **Where
found:** Wet, shady places. Canada south to Okla., Ala., Fla.
Flowers: June–Sept.
Use: Cooked green; remedy for Poison Ivy and nettles. See
Pale Touch-me-not, p. 78.   Early Spring (shoots);
                               Summer (leaves, stems)

DAY-LILY   **Young shoots, flower-**
Hemerocallis fulva   **Color pl. 7**   **buds, flowers, tubers**
Note the *unspotted* tawny blossoms (open 1 day only) facing
upward from the top of a *leafless* flowerstalk. Basal leaves light
green, *long, swordlike.* Root a tangle of small elongated tubers. 3–6 ft. (0.9–1.8 m). **Where found:** Waste ground, escaped
from gardens. Ontario to New Brunswick, south to Mo., Tenn.,
and N.C. **Flowers:** June–Aug.
Use: Salad, asparagus, cooked vegetable, fritters, seasoning. A
little-known but excellent food source. Add the early *shoots* to
salads or prepare like asparagus. Prepare the young *flowerbuds*
like green beans or, when older, like fritters. Use the fresh
flowers to make fritters, or the fresh, withered, or dried flowers
to season stews. Add crisp snow-white *tubers* found early in the
year to salads, or prepare like corn. Older, but still firm, tubers
can also be prepared like corn.   All Year (tubers);
    Early Spring (shoots); Summer (buds, flowers)

BUTTERFLY-WEED   **Poisonous**
Asclepias tuberosa
A showy milkweed with *erect,* flat-topped clusters of brilliant
orange flowers. Stems rough-hairy; juice watery, *not* milky.
Seedpods slender, 4–5 in. (10–12.5 cm) long. 1–2 ft. (30–60 cm).
Where found: Fields, dry soil. Minn., s. Ontario, s. Maine,
south to Texas and Fla. **Flowers:** May–Sept.
Warning: The stems and leaves may share the poisonous
qualities of the roots; do not use like Common Milkweed,
p. 112.

ORANGE

TURK'S-CAP LILY

SPOTTED TOUCH-ME-NOT

DAY-LILY

BUTTERFLY-WEED

X ½

MISCELLANEOUS PINK OR
PURPLE FLOWERS

WATER or PURPLE AVENS   Rootstock
Geum rivale   **Color pl. 6**
Note the irregular, compound basal leaves each with a *broad*
terminal segment. Upper leaves 3-lobed. Flowers often in clusters of 3; nodding, bell-like, brownish purple to yellowish.
Fruits hooked; in bristly heads. Rootstock reddish purple or
brownish. 1–2 ft. (30–60 cm). **Where found:** Wet places, wet
meadows, bogs. Canada south to Minn., n. Ill., Ohio, W. Va.,
Penn., and n. N.J. **Flowers:** May–Aug.
Use: Chocolatelike beverage. The boiled rootstocks make a
drink somewhat like hot chocolate. The flavor is much improved with the addition of milk and sugar.   ALL YEAR

EUROPEAN GREAT BURNET   Young leaves
Sanguisorba officinalis
Note the compound leaves with 7–15 pairs of *toothed, spade-
shaped leaflets.* Flowers with *4 petal-like sepals,* reddish-
purple; in dense oblong heads at the top of the stem. 10–24 in.
(25–60 cm). **Where found:** Wet meadows. Locally abundant,
Minn. to Maine. **Flowers:** June–Oct.
Use: Salad. The young leaves are a pleasant addition to fresh
salads, tasting mildly of cucumber.   EARLY SUMMER

ATAMASCO-LILY   Poisonous
Zephyranthes atamasco
Flower large, erect, 6-petaled, lilylike; white or pinkish, waxy-
looking. Leaves narrow, daffodil-like. Bulb suggests an onion,
but without the odor. 6–15 in. (15–37.5 cm). **Where found:**
Wet woods, clearings. Se. U.S., north to Va., in coastal plain or
lower Piedmont. **Flowers:** March–June.
Warning: The leaves are known to be poisonous to livestock.
The bulbs should be left alone.

PASSION-FLOWER, MAYPOP   Fruit
Passiflora incarnata
A weak, trailing or climbing vine with tendrils. Leaves finely
toothed, 3- (or 5-) lobed. The long-stalked showy flowers in the
upper leaf axils consist of 3–5 white sepals alternating with 3–5
white petals and overlaid by a sunburst structure (corona) of
purple or pink threads. Fruit a large, yellow, hen's-egg size
berry. To 25 ft. (7.5 m) long. **Where found:** Sandy soil; fields,
roadsides, thickets. S. Mo. to Va., south to Texas, Fla. **Flowers:** May–July. **Fruit:** July–Oct.
Use: Fresh fruit, cold drink, jelly. Although not very nutritious, the *ripe* fruit can be eaten fresh, or made into a cold drink
(simmer 5 min., strain, add lemon and sugar, chill) or jelly
(simmer, strain, add as much sugar as juice, add pectin).
                          LATE SUMMER–EARLY FALL

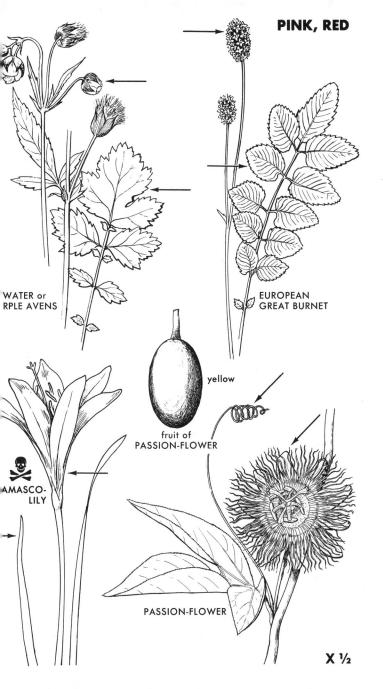

PINK, RED

WATER or
RPLE AVENS

EUROPEAN
GREAT BURNET

yellow

fruit of
PASSION-FLOWER

AMASCO-
LILY

PASSION-FLOWER

X ½

3 PETALS OR LOBES; BROAD LEAVES

TRILLIUMS   **Young leaves**
Trillium spp.
Flowers of the moist spring woods with broad leaves and solitary showy blossoms. *Leaves, petals, and sepals all in whorls of 3.*
Use: Salad, cooked green. Before fully unfolded, the young leaves are an excellent addition to salads, often tasting vaguely like raw sunflower seeds. The leaves can also be boiled for 10 min. and served with butter or vinegar. Once the flowers appear, the leaves become bitter. The berries and roots are inedible. **Note:** Although most trilliums should not be picked, the 3 species shown are occasionally abundant enough to allow their use.   EARLY SPRING
RED or PURPLE TRILLIUM, WAKEROBIN, *T. erectum.*
Color pl. 14. Flower maroon or purple (rarely white or greenish yellow), *foul smelling,* on an erect or reclining stalk. 7–16 in. (17.5–40 cm). **Where found:** Moist woods. Mich. to Quebec, south to Penn., n. Del.; in mts. to Ga. **Flowers:** April–June.
TOADSHADE, SESSILE TRILLIUM, *T. sessile.* Flowers *stalkless, erect;* maroon, purplish, or yellow-green (p. 146). Leaves often mottled. 4–12 in. (10–30 cm). **Where found:** Rich moist woods. Cen. Mo., w. N.Y., south to Ark., Ga. **Flowers:** April–June.
LARGE-FLOWERED or WHITE TRILLIUM, *T. grandiflorum.* White, turning pink with age. See p. 24.

WILD GINGER   **Rootstock**
Asarum canadense
A solitary, bell-shaped, red-brown flower with 3 spreading lobes sits on the ground between 2 woolly leafstalks. Leaves in pairs; large, heart-shaped. Rootstock smells of ginger. Colonial. 6–12 in. (15–30 cm). **Where found:** Rich, rocky woods. S. Canada south to Ark., n. Ga. **Flowers:** April–May.
Use: Candy, seasoning. The long horizontal rootstocks make an excellent candy when boiled until tender and then simmered in a rich sugar syrup (20–30 min.). Dried, they can be substituted for commercial ginger.   EARLY SPRING–FALL

WATER-SHIELD   **Young leaves, tubers**
Brassenia schreberi
Leaves 2–4 in. (5–10 cm) wide; oval, floating; greenish above purplish below. Stems slimy, submerged. Flowers small, dull purple, with 3 (or 4) petals. **Where found:** Quiet water. Most of our area. **Flowers:** June–Oct.
Use: Salad, cooked green, potato, flour. The newly emerging *leaves* can be added fresh to salads or prepared like spinach. The small *tubers* can be gathered like those of arrowheads (p. 24), and prepared like potatoes, or ground into flour.
         SPRING (leaves); FALL–EARLY SPRING (tubers)

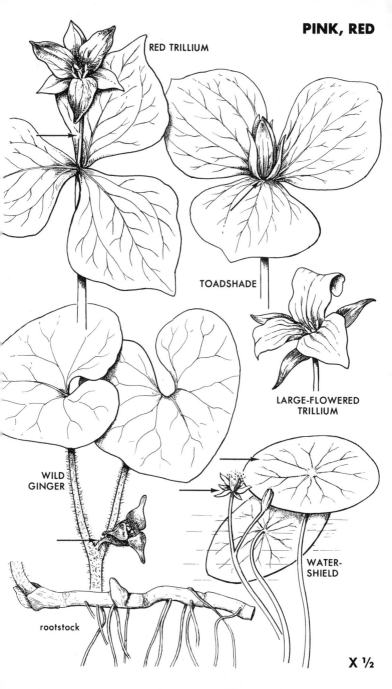

PINK, RED

RED TRILLIUM

TOADSHADE

LARGE-FLOWERED
TRILLIUM

WILD
GINGER

WATER-
SHIELD

rootstock

X ½

SHOWY 4-PETALED FLOWERS

RIVER-BEAUTY   Shoots, young leaves
Epilobium latifolium

A low, matted, arctic relative of Fireweed (below) with numerous stems arching upward from the same base. Flowers in upper leaf axils; magenta; up to 2 in. (5 cm) across. Leaves toothless, elliptic-ovate to lance-shaped; fleshy, often powdered with white. 4–16 in. (10–40 cm). **Where found:** Gravelly streambeds, shores. Gaspé Peninsula northward. **Flowers:** July–Aug.

Use: Asparagus, cooked green. The new *shoots* and tender *young leaves* are much superior to those of Fireweed; prepare them similarly.   SPRING–EARLY SUMMER

FIREWEED   Young shoots, leaves
Epilobium angustifolium   **Color pl. 9**

A tall ruddy-stemmed herb with a large showy spike of 4-petaled rose-purple flowers, drooping flowerbuds, and reddish seedpods angling upward. Leaves usually toothless, alternate. 3–7 ft. (0.9–2.1 m). **Where found:** Newly cleared woods, fire-desolated areas. Subarctic south to Iowa, Ind., n. Ohio, Md.; in mts. to Ga. **Flowers:** July–Sept.

Use: Asparagus, cooked green, tea. The *young shoots* can be prepared like asparagus and the tender *young leaves* like spinach. The taste may become bitter and unpalatable as the plant grows older. The *mature leaves* can be dried and used to make tea.   SPRING (shoots); SUMMER (leaves)

VIRGINIA MEADOW-BEAUTY   Tender leaves, tubers
Rhexia virginica   **Color pl. 2**

Note the 4 showy rose or magenta petals and slender *golden anthers.* Leaves opposite, broad at base; stem squarish, with 4 thin lengthwise ridges; slightly hairy. Root usually with 1 or 2 small tubers. Often in dense colonies. 1–1½ ft. (30–45 cm). A similar species, *R. mariana* (not shown), is pale pink, hairier, with narrower leaves tapering into slender bases; root without tuber. **Where found:** Wet or moist sandy soil. S. Ontario, Nova Scotia, south to Mo., Tenn., Ala., and Ga. **Flowers:** June–Sept.

Use: Salad, cooked green, nibble. The tender *leaves,* tasting slightly sweet-sour, are a fine addition to salads. They can also be boiled for about 10 min. and served with lemon or vinegar. The firm *tubers* can be chopped and added to salads, or eaten as a pleasant nutlike nibble.   SUMMER

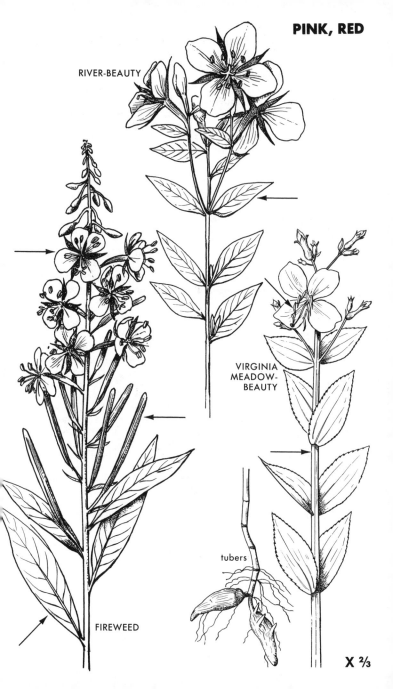

PINK, RED

RIVER-BEAUTY

VIRGINIA
MEADOW-
BEAUTY

FIREWEED

tubers

X ⅔

4 PETALS; TERMINAL CLUSTERS: MUSTARDS

 CUCKOO-FLOWER, LADY'S-SMOCK   Young leaves
Cardamine pratensis
Note the *numerous small paired leaflets;* roundish on basal leaves, narrow on upper leaves. Basal leaves in rosettes. Flowers white or pale rose. 8–20 in. (20–50 cm). **Where found:** Swamps, springs, wet meadows, wet woods. N. Canada to Minn., n. Ill., W. Va., and n. N.J. **Flowers:** April–June.
Use: Salad. Pungent; excellent.   EARLY SPRING

 **SEA-ROCKET**   Leaves, stems, flowerbuds,
Cakile edentula   young pods
A common seaside plant with *succulent* stems and fleshy, peppery-pungent leaves. Flowers small, pale purple. Seedpods distinctive (inset). 6–12 in. (15–30 cm). A similar species, *C. lanceolata* (not shown), found on beaches in Fla., has narrower, more deeply cleft leaves. **Where found:** Beaches, seacoast. Coast from s. Labrador south to S.C. **Flowers:** June–Sept.
Use: Salad, cooked vegetable. The *leaves* and *young seedpods* can be added to salads. The *flowerbuds, green pods, tender stems, and leaves* can be boiled for 5–10 min.   SUMMER

TOOTHWORTS   Rootstock
Dentaria spp.
Low woodland plants of spring with small but quite visible clusters of 4-petaled, white to pale pink (when fading) flowers. A horizontal, whitish rootstock is shallowly buried; pungent to taste. Colonial. 3 typical species shown.
Use: Horseradishlike condiment, salad. The crisp rootstocks are peppery-pungent; mixed with vinegar and a pinch of salt, they can be substituted for horseradish; chopped up, they make a lively addition to salads.   EARLY SPRING–SPRING
TOOTHWORT, PEPPERWORT, *D. diphylla.* **Color pl. 14.**
2 nearly opposite stem leaves and long-stalked basal leaves. Each leaf with 3 broad, coarsely blunt-toothed leaflets. 8–14 in. (20–35 cm). **Where found:** Moist woods. S. Canada south to Ky., S.C. **Flowers:** April–June.
CUT-LEAVED TOOTHWORT, *D. laciniata.* Leaves in a *whorl of 3,* each with *3 long, narrow, jagged-lobed segments.* Basal leaves usually absent. Flowers white or pale lavender. 8–15 in. (20–37.5 cm). **Where found:** Rich moist woods. Minn. to Vt., south to La., Fla. **Flowers:** March–May.
LARGE TOOTHWORT, *D. maxima.* Resembles *D. diphylla,* but with *3 leaves* growing *at different levels* along the stem. Flowers white or pale purple. 6–16 in. (15–40 cm). **Where found:** Woods, streambanks. Wisc. to s. Maine, south to Tenn., W. Va., Penn., and s. New England. **Flowers:** April–June.

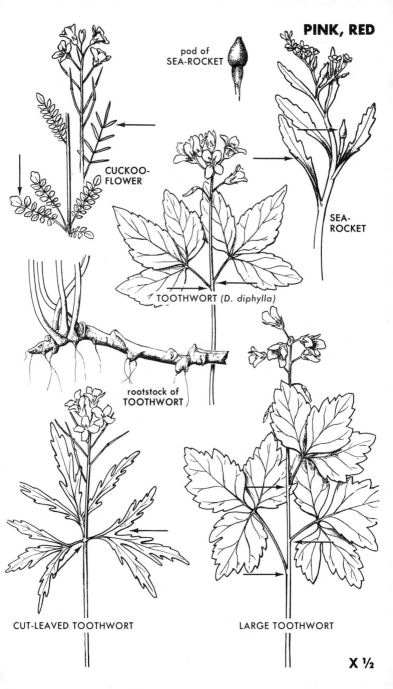

PINK, RED

pod of
SEA-ROCKET

CUCKOO-
FLOWER

SEA-
ROCKET

TOOTHWORT *(D. diphylla)*

rootstock of
TOOTHWORT

CUT-LEAVED TOOTHWORT

LARGE TOOTHWORT

X ½

CREEPING OR MATTED EVERGREEN PLANTS

CRANBERRIES   **Fruit**
Vaccinium spp.
Low creeping shrubs with slender stems and small, alternate, oval, evergreen leaves. Fruit red, juicy. 3 species shown.
Use: Cooked fruit, jelly, drink. See p. 222.   FALL–WINTER
LARGE CRANBERRY, *V. macrocarpon.* **Color pl. 4.** Recurved petals form a "Turk's cap"; stamens form a "beak." Leaves blunt; flowerstalks do *not* spring from tip of stem. **Where found:** Open bogs. S. Canada south to Ill., Va. **Flowers:** June–Aug. **Fruit:** Sept.–Nov. (often overwintering).
SMALL CRANBERRY, *V. oxycoccus.* Smaller than Large Cranberry. Leaves *pointed, white beneath, with rolled edges.* Flowerstalks spring from tip of stem. **Where found:** Bogs, wet peaty soil. Canada south to Minn., Mich., n. Ohio, N.J.; in mts. to N.C. **Flowers:** May–July. **Fruit:** Aug.–Oct. (or later).
MOUNTAIN CRANBERRY, *V. vitis-idaea.* Flowers bell-like, without recurved lobes. Leaves blunt, *dotted with black beneath.* 2–7 in. (5–17.5 cm). **Where found:** Rocky or dry peaty soil. Canada south to n. Minn., n. New England. **Flowers:** June–July. **Fruit:** Aug.–Oct. (often remain through winter).

BLACK CROWBERRY   **Fruit**
Empetrum nigrum
Low, mat-forming, with tiny pink flowers tucked in the axils of *short, needlelike* leaves. Fruit a juicy black berry. **Where found:** Open peaty soil. Arctic south to northern edge of U.S. **Flowers:** July–Aug. **Fruit:** July–Nov. (often overwintering). **Use:** Fresh or cooked fruit. The flavor of the berries improves after freezing or when cooked with a little lemon juice and served with cream and sugar. Add the fresh berries to pancakes, muffins, and the like.   LATE SUMMER–WINTER

PARTRIDGEBERRY   **Fruit**
Mitchella repens   **Color pl. 15**
The *twin,* pink or white, 4-petaled flowers terminating the creeping stem produce a solitary dry red berry. Leaves in pairs, roundish. **Where found:** Moist woods. Much of our area. **Flowers:** May–July. **Fruit:** July–winter.
Use: Nibble, salad. See pp. 36, 174.   FALL–EARLY SPRING

TRAILING ARBUTUS   **Corolla (flower tube)**
Epigaea repens
A trailing evergreen shrub with leathery oval leaves and woody brown-hairy stems. Flowers pink or white, clustered, with 5 lobes flaring from a short tube. **Where found:** Dry soil, woods, clearings. Canada, n. U.S.; south in uplands and mts. to Ala. and Ga. **Flowers:** Late Feb.–May.
Use: Nibble, salad. See pp. 36, 224.   EARLY SPRING

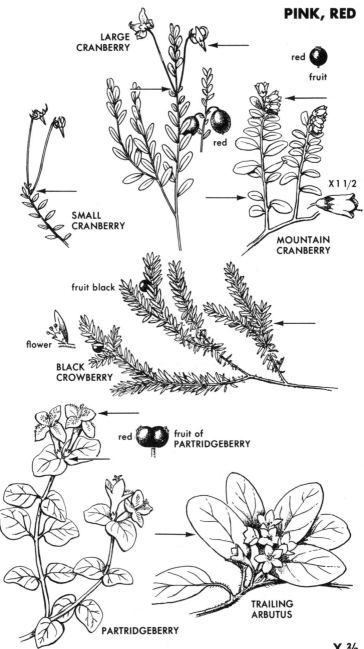

PINK, RED

LARGE
CRANBERRY

red
fruit

red

SMALL
CRANBERRY

X 1 1/2

MOUNTAIN
CRANBERRY

fruit black

flower

BLACK
CROWBERRY

red   fruit of
PARTRIDGEBERRY

PARTRIDGEBERRY

TRAILING
ARBUTUS

X 2/3

5 PETALS; LOW FLOWERS

SPRING-BEAUTIES   **Corm**
Claytonia spp.
A single pair of smooth leaves are attached halfway up the delicate stems. Petals white to pale pink, with darker pink veins. Root a small — $\frac{1}{4}$-$1\frac{1}{2}$ in. (0.6–3.8 cm) — corm; buried 3–5 in. (7.5–12.5 cm). Colonial. 6–12 in. (15–30 cm).
Use: Potato. Although time consuming to dig up in quantity, the small corms are excellent boiled for 10–15 min., stripped of their tough outer jackets, and served with butter. They can also be baked, fried, or mashed. **Note:** Collect only when abundant.   Early Spring–Spring
SPRING-BEAUTY, *C. virginica.* Narrow, linear leaves. **Where found:** Moist woods, rich soil. Minn. to s. Quebec, south to Texas, Ala., and Ga. **Flowers:** March–May.
CAROLINA SPRING-BEAUTY, *C. caroliniana.* **Color pl. 14.** Leaves *much wider,* slender-stalked. **Where found:** Woods, uplands. S. Canada south through Appalachians and westward. **Flowers:** March–May.

STORKSBILL, ALFILARIA   **Young leaves**
Erodium cicutarium
Leaves *twice-cut, fernlike;* soft-hairy; often in prostrate rosettes. Flowers less then $\frac{1}{2}$ in. (1.3 cm) across, rose-colored. Seedpods beaklike, stem reflexed. 3–12 in. (7.5–30 cm). **Where found:** Dry sandy soil, fields, roadsides. Mich. to Quebec, south to Texas, Ark., Tenn., Va. **Flowers:** March–Nov.
Use: Salad, cooked green. The tender young leaves can be added fresh to salads, or boiled for 10 min. and served with butter.   Early Spring

WOOD-SORRELS   **Leaves**
Oxalis spp.
Low, delicate, woodland flowers. Leaves cloverlike, with 3 inversely heart-shaped leaflets that often fold along a central crease. *Note sour taste.* See p. 72 for yellow wood-sorrels.
Use: Salad, cold drink. The fresh leaves make an excellent sour addition to salads. To make a refreshing drink, steep the leaves for 10 min. in hot water, chill, and add sugar or honey. Rich in vitamin C. **Warning:** Excessive consumption over an extended period of time may inhibit the absorption of calcium by the body.   Spring–Summer
VIOLET WOOD-SORREL, *O. violacea.* Note the 5 flaring *rose-purple or purplish violet* petals. 4–8 in. (10–20 cm). **Where found:** Open woods, banks, prairies. Minn., Ind., Ohio, N.Y., Mass., south to Texas, Fla. **Flowers:** April–July.
COMMON WOOD-SORREL, *O. montana.* **Color pl. 15.** Petals white or pale pink with prominent pink veins. 3–5 in. (7.5–12.5 cm). **Where found:** Cool, moist woods. Canada, northern edge of U.S.; south in mts. to Tenn. and N.C. **Flowers:** May–Aug.

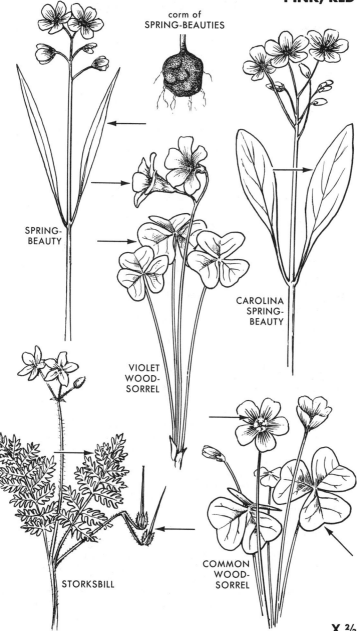

corm of
SPRING-BEAUTIES

SPRING-
BEAUTY

CAROLINA
SPRING-
BEAUTY

VIOLET
WOOD-
SORREL

STORKSBILL

COMMON
WOOD-
SORREL

X ⅔

5 SHOWY ROSELIKE PETALS

ROSES   **Petals, fruit (hips);**
Rosa spp.   **leaves (Sweetbrier)**

A large and familiar group of bristly or thorny shrubs with showy, 5-petaled, pink or deep rose flowers. Fruit bright red, with 5 prominent calyx lobes at end. Numerous species in our area, all equally edible; 3 shown.

Use: Jam, tea, candy, emergency food. An excellent jam can be made with rose *hips* and sour apples. Discard the stems and calyx lobes, and combine the hips with sliced apples (3 parts hips to 1 part apples); add just enough water to cover, and simmer until soft. Strain the mixture through cheesecloth, sweeten to taste and boil hard until ready to set (about 15 min.). Commercial pectin must be added if the sour apples are omitted. Or syrup can be made by leaving out the pectin. Fresh or dried hips, fresh *petals,* and *Sweetbrier leaves* can be steeped in hot water for 10 min. to make tea. The fresh petals can be added to salads, made into jelly, or candied. The pulpy exterior of the hips can be eaten raw. Because rose hips are often held on the bush through winter, they make an excellent survival food. Rich in vitamin C.   SUMMER (petals);
LATE SUMMER–EARLY FALL (fruit)

WRINKLED ROSE, *R. rugosa.* **Color pl. 1.** A coarse species forming dense clumps. Stems stout, *bristly;* leaves usually with 7–9 conspicuously *wrinkled* leaflets. Flowers large, deep rose to white. Fruit sizable. 2–6 ft. (0.6–1.8 m). **Where found:** Sand dunes, seashores, roadsides. Minn. through Great Lakes region to Quebec and New Brunswick, south along the coast to N.J. **Flowers:** June–Sept.

SWEETBRIER, *R. eglanteria.* The long, arching main stems are clustered with numerous smallish pink blossoms. Leaves smell somewhat of green apples when crushed. Leaflets roundish, *double-toothed.* 4–6 ft. (1.2–1.8 m). **Where found:** Roadsides, clearings. Throughout. **Flowers:** May–July.

PASTURE or CAROLINA ROSE, *R. carolina.* A low, slender rose with *slender, straight* (*not recurved*) *thorns.* 1–3 ft. (30–90 cm). **Where found:** Sandy or rocky pastures, dry open woods. Minn., Wisc., Mich., s. Ontario, N.Y., and New England, south to Texas and Fla. **Flowers:** May–July.

PURPLE-FLOWERING RASPBERRY   **Fruit**
Rubus odoratus   **Color pl. 12**

A straggling shrub with large *maplelike* leaves and thornless stems covered with sticky reddish hairs. Flowers rose-purple; 1–2 in. (2.5–5 cm) across. Berries dull red, somewhat flattened, cup-shaped. 3–6 ft. (0.9–1.8 m). **Where found:** Moist rocky woods, ravines, thickets. S. Canada south in mts. to Tenn., Ga. **Flowers:** June–Sept. **Fruit:** July–Sept.

Use: Fresh or cooked fruit. The berries are rather tart and dry, but quite edible. See other raspberries, p. 184.   SUMMER

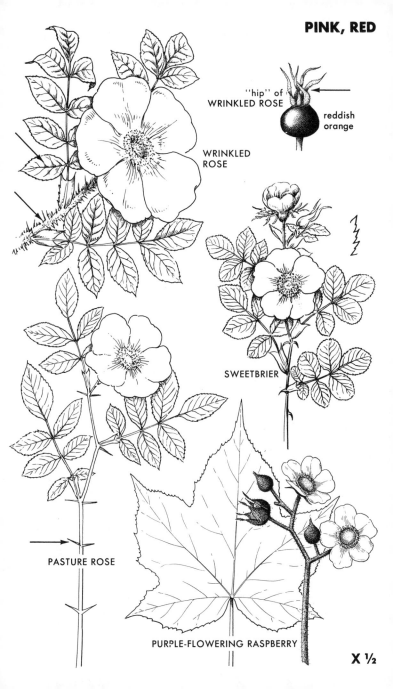

PINK, RED

"hip" of
WRINKLED ROSE
reddish orange

WRINKLED ROSE

SWEETBRIER

PASTURE ROSE

PURPLE-FLOWERING RASPBERRY

X ½

5 PETALS; STAMENS IN
BUSHY COLUMNS

MALLOWS   **Young leaves; fruit (Cheeses)**
Malva spp.
The showy, 5-petaled, pink or pale lavender petals and bushy
column of stamens suggest a small hibiscus. One or more
species found throughout our area; all are potentially edible; 3
shown.
Use: Okralike thickener, cooked green; salad (Cheeses). The
tender *young leaves* can be used okralike to thicken soups and
stews, or they can be boiled for 10–15 min. and served with
butter. The tender green *fruit of Cheeses* are a pleasant nibble
or addition to salads.   SPRING (leaves);
                    SUMMER (leaves, fruit of Cheeses)
MUSK MALLOW, *M. moschata.* Note the notched, pink-lavender or white flowers and deeply cut leaves. 1–2 ft. (30–
60 cm). **Where found:** Fields, roadsides. Ontario, Newfoundland, south to Tenn., Md., and Del. **Flowers:** June–Sept.
CHEESES, COMMON MALLOW, *M. neglecta.* **Color pl. 8.**
A small creeping or trailing weed. Flowers in leaf axils, pale
rose-lavender or white with magenta veins, ½ in. (1.3 cm)
across; petals notched. Leaves roundish, with 5–7 scalloped
lobes. The name refers to the cheeselike form of the fruit.
Where found: Barnyards, waste places. Throughout. **Flowers:** April–Oct.
HIGH MALLOW, *M. sylvestris.* Note the pink or lavender
petals *strongly veined with dark red.* Leaves broad, bluntly
5-lobed. 1–3 ft. (30–90 cm). **Where found:** Roadsides. Local,
but widespread. **Flowers:** May–July.

MARSH MALLOW   **Young leaves, flowerbuds,**
Althaea officinalis   **Color pl. 1**   **roots**
An erect marsh plant growing in soft-stemmed clumps. Leaves
gray-green, velvety-textured; 1–4 in. (2.5–10 cm) long. Upper
leaves coarsely toothed; lower leaves heart-shaped or 3-lobed.
Flowers pale pink, 1–1½ in. (2.5–3.8 cm) across. Root large,
white, mucilaginous. 2–4 ft. (0.6–1.2 m). **Where found:** Edges
of saltmarshes. Coast from Conn. to Va. **Flowers:** Aug.–Oct.
Use: Candy, cooked vegetable, okralike thickener, pickled capers. The original campfire marshmallows were made from the
roots of this plant: Strip off the corky outer layer, cut into
small pieces, and boil with sugar until very thick; strain out the
roots and drop the candy on wax paper to cool. The peeled
roots also make an interesting cooked vegetable: Slice into thin
sections, boil for 20 min., and fry with butter and onions until
golden brown. The *young leaves* can be used like the mallows
(above) to thicken soups and stews. The *flowerbuds* can be
pickled and used as a substitute for capers.
                    EARLY SUMMER (leaves); SUMMER (flowerbuds);
                    FALL–EARLY SPRING (roots)

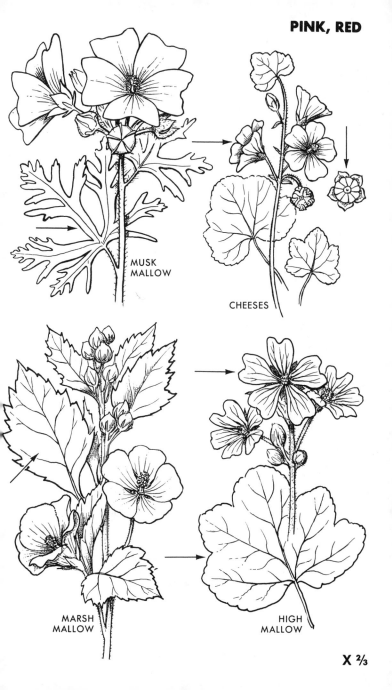

MUSK MALLOW

CHEESES

MARSH MALLOW

HIGH MALLOW

X ⅔

5-PART FLOWERS: MISCELLANEOUS

COMFREY   **Young leaves**
Symphytum officinale
A rough-hairy, branching perennial. Leaves large, alternate, spear-shaped, tapering into a *winged* stem. Flowers 5-lobed, bell-like; in a 1-sided curling flower cluster with a *pair of winglike leaves;* white, yellow, pink, or purple, fading to blue. 2–3 ft. (60–90 cm). **Where found:** Damp roadsides, waste places. **Flowers:** June–Sept.
Use: Cooked green, tea. The newly emerging leaves of March or April are excellent prepared like spinach; the hairiness is reduced upon cooking. Slightly older leaves are bitter but still quite palatable if boiled in several changes of water. Steep the dried leaves to make tea.   EARLY SPRING–SPRING

SEA-PURSLANE   **Leaves and stems**
Sesuvium maritimum
A mostly prostrate, branching annual, with succulent stems and *fleshy, paddle-shaped* leaves. Flowers tiny, petal-less, tucked in leaf axils; greenish without, purplish within. 2–12 in. (5–30 cm). A taller, coarser species, *S. portulacastrum* (not shown), is found on coastal sands from N.C. to Miss. **Where found:** Damp sand. Coastal shores, Long Island south to Fla. and Texas. **Flowers:** May–Sept.
Use: Cooked green. Boil the tender leaves and stems for 10–15 min., changing the water 1 or 2 times to get rid of excess salt. Serve with butter.   EARLY SUMMER

DOGBANES   **Poisonous**
Apocynum spp.
Branching plants with short-stalked, paired, ovate leaves. Stems reddish, with *milky juice.* Flowers numerous, pale pink, nodding, bell-like. Seedpods in pairs, slender, 3–8 in. (7.5–20 cm) long. 1–4 ft. (0.3–1.2 m).
Warning: Do not mistake the slender shoots for the stouter ones of Common Milkweed, p. 112. Fortunately, young dogbane shoots are hairless and quickly branching.
SPREADING DOGBANE, *A. androsaemifolium.* Flowers with strongly recurved lobes and deep-rose stripes inside. **Where found:** Thickets, roadsides. Canada south to Ark., Ill., cen. Ind., Ohio, W. Va., and uplands of N.C. **Flowers:** June–July.
INTERMEDIATE DOGBANE, *A. medium.* Hybrid between Spreading Dogbane and Indian Hemp (p. 48). Flowers lack recurved lobes; less pink, often white. **Where found:** Fields, roadsides, shores. S. Canada south to Texas, Mo., Tenn., and Va. **Flowers:** June–Aug.

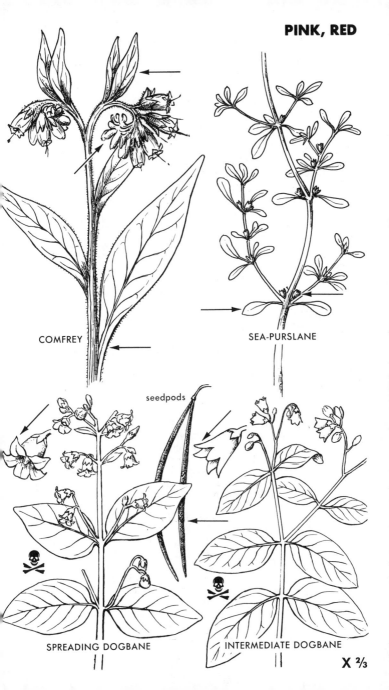

COMFREY

SEA-PURSLANE

seedpods

SPREADING DOGBANE

INTERMEDIATE DOGBANE

X ⅔

DOMED, UMBRELLALIKE CLUSTERS

 ORPINE, LIVE-FOREVER   **Young leaves, tubers**
Sedum purpureum

 Leaves light green, succulent, coarsely-toothed; alternate or in whorls of 3. Stems stout, fleshy. Flowers 5-petaled, pink-purple. Root a mass of fingerlike tubers. 12–30 in. (30–75 cm). **Where found:** Roadside banks, field edges, waste places. S. Canada south to Ind., Md. **Flowers:** July–Sept.

 Use: Salad, cooked green, cooked vegetable, pickle. The *young leaves* can be added to salads or boiled for 5–10 min. The crisp *tubers* can either be boiled for 20 min. and served with butter, or pickled in seasoned vinegar.

 SPRING (leaves); FALL–EARLY SPRING (tubers)

ROSEROOT   **Leaves and stems**
Sedum rosea

Note the *spirally overlapping,* succulent leaves. The *4-petaled* flowers may be either yellowish (staminate) or purplish (pistillate). 5–10 in. (13–25 cm). **Where found:** Rocks. Arctic south in mts. to Penn., on coast to Maine. **Flowers:** May–Aug.
Use: Salad, cooked green. See Orpine (above).   SUMMER

COMMON MILKWEED   **Young shoots, leaves,**
Asclepias syriaca   **unopened flowerbuds,**
Color pl. 10   **flowers, young pods**

Stout, *downy,* gray-green, usually unbranched. Note the milky juice of the broken stems, and the unique flower structure (inset). Flowers in domed, often somewhat drooping, clusters in leaf axils; greenish purple to dull purple, buff, or whitish; often fragrant. Seedpods pointed, gray-green, *warty.* 3–5 ft. (0.9–1.5 m). **Where found:** Dry soil, fields, roadsides. S. Canada south to Kans., Tenn., Ga. (uplands). **Flowers:** June–Aug.
Use: Asparagus, cooked green, cooked vegetable, fritters. The milky juice of the broken stems and leaves is bitter and mildly toxic. Fortunately, both of these properties are dispelled upon boiling, and Milkweed becomes one of the better wild vegetables. Cover the young *shoots,* up to 6 in. (15 cm), with boiling water and cook for 15 min., using several changes of water. The first few changes of water should be fairly rapid with just over a minute between each change; be sure to use boiling water when making each change, as covering the plants with cold water and bringing them to a boil tends to fix their bitter flavor. The tender young *top leaves, flowerbuds,* and small hard *young pods* are all prepared in much the same way as the shoots. The *flowers* can be dipped into boiling water for 1 min., covered with batter, and fried to make fritters. **Caution:** Do not confuse young shoots with those of dogbanes (pp. 48, 110) or Butterfly-weed (p. 92). Common Milkweed shoots downy-hairy, with milky sap; dogbanes hairless; Butterfly-weed lacks milky sap.
SPRING (shoots, leaves); SUMMER (buds, flowers, pods)

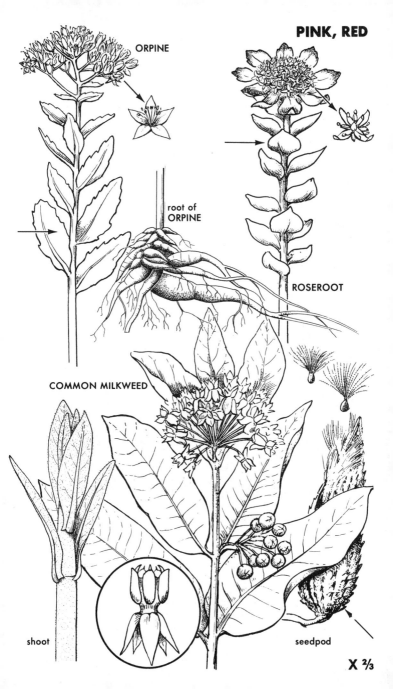

ORPINE

PINK, RED

root of ORPINE

ROSEROOT

COMMON MILKWEED

shoot

seedpod

X ⅔

DOMED CLUSTERS; 6 PETALS
GRASSLIKE LEAVES: ONIONS

WILD ONIONS   Leaves, underground bulb;
Allium spp.   bulblets (Wild Garlic)

Widespread, familiar plants with grasslike basal leaves and small, 6-petaled flowers. Note the *odor of onions*. All species edible; 4 shown. See also Wild Leek, p. 52.

Use: Cooked vegetable, pickle, salad, seasoning, cooked green. The underground *bulbs* are excellent boiled, pickled, added to salads, or used as a seasoning. The tender *leaves* (before the flowerstalks appear) can be cooked as greens along with the bulbs, or added raw to salads. The green *bulblets* that form after the flowers bloom on Wild Garlic can be made into outstanding pickles. Do not confuse Field Garlic with Wild Garlic; although its parts are quite edible, they cause a lingering odor similar to that of commercial garlic.

SPRING (leaves); SUMMER (bulblets); ALL YEAR (bulbs)

NODDING WILD ONION, *A. cernuum.* Because of the *bend* at the tip of the stem, the pink or white flower cluster droops or nods. 1–2 ft. (30–60 cm). **Where found:** Rocky soil, open woods, slopes. Minn., Mich., N.Y. south to upland Texas, Miss., Ala., and Ga. **Flowers:** July–Aug.

WILD ONION, *A. stellatum.* Somewhat resembles the preceding species, but the stem is straight. Flowers pink. 1–2 ft. (30–60 cm). **Where found:** Open rocky places, prairies. Saskatchewan to Mo., east to Ohio. **Flowers:** July–Aug.

FIELD GARLIC, *A. vineale.* Similar to Wild Garlic (below), but flowers more numerous. Note the *single* spathe below the flower cluster and the round, *hollow* leaves rising part way up the stem. 1–3 ft. (30–90 cm). **Where found:** Fields, roadsides. Ill. to Mass., south to Ark., Ga. **Flowers:** May–July.

WILD GARLIC, *A. canadense.* The sparse pink or white flowers are mixed with bulblets, or when absent, are replaced by bulblets with "tails." Note the *3-parted* spathe below the flower cluster. Leaves mostly basal, *not hollow;* flattened. 8–24 in. (20–60 cm). **Where found:** Meadows, clearings. Minn., Wisc., s. Ontario, s. Quebec south to Texas and Fla. **Flowers:** April–July.

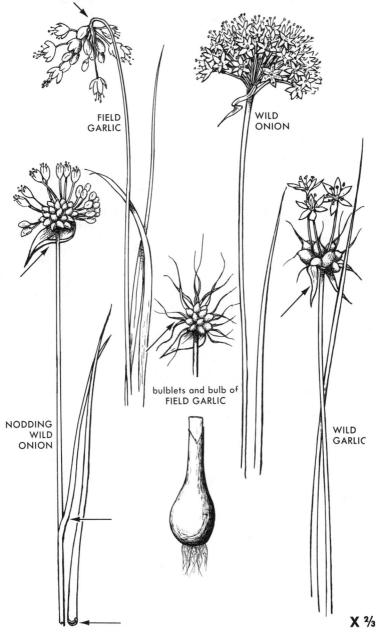

FIELD GARLIC

WILD ONION

NODDING WILD ONION

bulblets and bulb of FIELD GARLIC

WILD GARLIC

X ⅔

SLENDER CLUSTERS OF TINY FLOWERS

LADY'S-THUMB, REDLEG   **Young leaves**
Polygonum persicaria
An erect or sprawling weed with spikelike clusters of tiny pink flowers and narrow leaves that often show a *dark triangular "thumbprint."* Mature stems *reddish,* jointed, with a papery *fringed sheath* at each joint. 6–24 in. (15–60 cm). **Where found:** Waste places. Throughout. **Flowers:** June–Oct.
Use: Cooked green, salad. The young leaves make an acceptable wild spinach when boiled for 5–10 min. and served with vinegar; raw they can be added to salads. **Note:** Lady's-thumb belongs to a widespread, distinctive group of nonpoisonous plants known as the smartweeds, *Polygonum* spp. Although some species in this group are too acrid-peppery to be eaten, many are mild tasting and could potentially be used in the same way as Lady's-thumb.   Spring

ALPINE SMARTWEED, ALPINE BISTORT
Polygonum viviparum   **Rootstock, bulblets**
Color pl. 4
A small erect arctic perennial topped by a slender cluster of pale pink or white flowers; small, purple or reddish *bulblets* usually replace bottom flowers. Stems with *slender sheaths* at the base of the narrow, dark green leaves. Basal leaves long-stalked. Rootstock small, fleshy, tuberlike. 4–8 in. (10–20 cm). **Where found:** Damp limy or rocky soil. Arctic south to n. Minn., n. Mich., and mts. of Quebec and n. New England. **Flowers:** July–Aug.
Use: Nibble, cooked vegetable. The tiny *bulblets,* stripped from the flowerheads, make a pleasant nutty nibble. The short *rootstocks* can be eaten raw or boiled, but are best roasted; they do not need to be peeled before being used. An excellent alpine or arctic survival food.   Summer–Fall

SHEEP or COMMON SORREL   **Tender leaves**
Rumex acetosella   **and stems**
Note the spreading lobes and *sour taste* of the small *arrow-shaped* leaves. Flowers tiny in sparse, branching spikes; reddish or greenish. 4–12 in. (10–30 cm). See also p. 154. **Where found:** Poor soil; thin fields, roadsides. Throughout. **Flowers:** June–Oct.

Use: Salad, cooked green, cold drink. Add to salads, or boil for 3–5 min. and serve with butter; steep for 10–15 min. in hot water and serve the chilled and sweetened juice as lemonade. Excellent; rich in vitamin C. **Caution:** See wood-sorrels, p. 104.   Spring–Summer

116

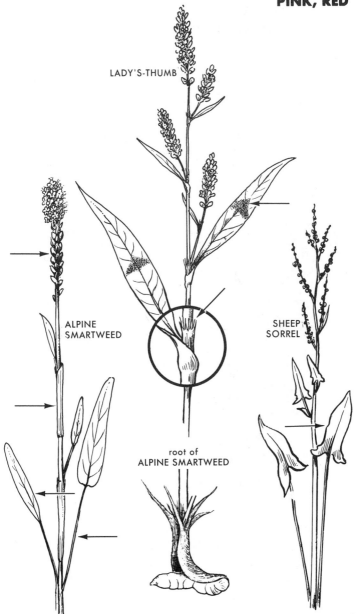

PINK, RED

LADY'S-THUMB

ALPINE
SMARTWEED

SHEEP
SORREL

root of
ALPINE SMARTWEED

X ⅔

PEPPERMINT   Leaves
Mentha piperita   **Color pl. 6**

Note the *peppermint odor* and *hot taste* of the smooth, short-stalked leaves. Stems purplish, branching. Flowers pink to pale violet, in short *or interrupted spikes.* 1½–3 ft. (45–90 cm). See other mints, p. 138. **Where found:** Brooksides, wet meadows, roadside ditches. Throughout. **Flowers:** June–Oct.

Use: Tea, flavoring. The fresh or dried leaves make an excellent tea when steeped in hot water for 5–10 min. and sweetened to taste. They can also be used to flavor jellies, sauces, dressings, and drinks.   SUMMER

CURLED MINT   Leaves
Mentha crispa

Note the broad aromatic leaves with the *ragged-toothed, crisped* margins. The pinkish flowers are in slender terminal clusters. 1½–3 ft. (45–90 cm). **Where found:** Wet spots, roadside ditches. Mich. to Mass., south to Penn., N.J. **Flowers:** June–Sept.

Use: Tea, flavoring. See Peppermint (above).   SUMMER

BERGAMOTS   Leaves and flowerheads
Monarda spp.

Bergamots are large coarse mints with showy flowerheads, opposite leaves, and square stems. Crushed leaves aromatic, minty. 2 species shown.

Use: Tea. Steep the fresh or dried leaves and flowerheads in hot water for about 10 min. and sweeten to taste. Excellent mixed with other teas.   SUMMER

BEE-BALM, OSWEGO-TEA, *M. didyma.* The large, *ragged head* of bright *scarlet* flowers and reddish bracts are distinctive. 2–3 ft. (60–90 cm). **Where found:** Moist woods and thickets, streambanks. Mich. to N.Y., south, chiefly in uplands, to Tenn. and Ga. **Flowers:** Late June–Sept.

WILD BERGAMOT, *M. fistulosa.* Our most abundant species. *Pinkish or pale lavender,* with bracts commonly tinged with lilac. *M. clinopardia* (not shown) is a paler species, with whitish flowerheads and bracts. 2–3 ft. (60–90 cm). **Where found:** Dry edges, thickets, clearings. Minn. to w. New England, south (chiefly in uplands) to e. Texas, Ga. **Flowers:** July–Sept.

PURPLE BERGAMOT, *M. media* (not shown). Resembles Wild Bergamot, but with dark *rose-purple* flowers and purplish bracts. **Where found:** Roadside thickets, often cultivated. Ind., Ontario, N.Y., south (in mts.) to Tenn. and N.C.

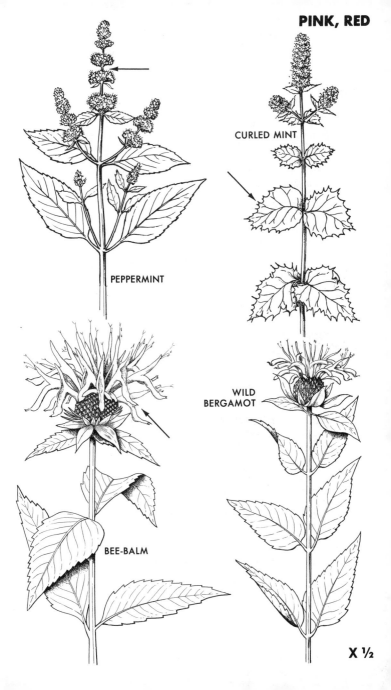

PEPPERMINT

CURLED MINT

BEE-BALM

WILD BERGAMOT

X ½

SQUARE STEMS; LIPPED FLOWERS; CLUSTERS IN LEAF AXILS: MINTS (2)

BASIL **Leaves**
Satureja vulgaris
A downy-hairy plant with toothed, pointed-oval leaves and pink-purple flowers. White hairs on the calyx and bracts make the flower clusters appear *woolly.* 9–18 in. (22.5–45). **Where found:** Roadsides, pastures, edges of woods, shores. Canada, n. U.S.; in uplands to Tenn. and N.C. **Flowers:** June–Sept. **Use:** Seasoning, tea. Although somewhat mild, the dried leaves can be used in lieu of commercial basil. The fresh leaves can be made into tea. SUMMER

WOUNDWORT **Tubers**
Stachys palustris
A downy plant with hairy stems and lance-shaped, stalkless or short-stalked leaves. Flowers in a whorled spike; pale pink or purple, with darker purple spots. Calyx tubes green or maroon, downy. 2–3 ft. (60–90 cm). **Where found:** Wet meadows, ditches, waste places, shores. Canada, n. U.S. **Flowers:** July–Sept.
Use: Salad, cooked vegetable, pickle. See Hyssop Hedge-nettle (below). FALL–EARLY SPRING

DITTANY **Leaves**
Cunila origanoides
A highly aromatic plant with wiry, much-branched stems and stalkless leaves. Leaves hairless, dotted with clear spots. Flowers in tufts, their 5 lobes nearly equal. Note the *2 long protruding stamens* and pistil. 8–16 in. (20–40 cm). **Where found:** Dry open woods, clearings. Okla., Mo., Ill., Ind., Ohio, e. Penn., and se. N.Y., south to Texas and Fla. **Flowers:** July–Oct.
Use: Tea. Steep the fresh or dried leaves for 10 min. in hot water and sweeten. Refreshing. SUMMER

HYSSOP HEDGE-NETTLE **Tubers**
Stachys hyssopifolia
A smooth erect plant often branching from the base. Leaves *narrow,* stalkless, usually *toothless.* Flowers pink, mottled with purple and white. 4–30 in. (10–75 cm). **Where found:** Moist or wet, usually sandy soil; shores, meadows. Se. Mass. and Conn. south along coast to n. Fla.; locally inland from e. Mo. to Penn. and south. **Flowers:** June–Oct.
Use: Salad, cooked vegetable, pickle. The sandy soil around the dry, shriveled stems is often laced with crisp, white, elongated tubers. These are excellent raw, cooked, or pickled.
FALL–EARLY SPRING

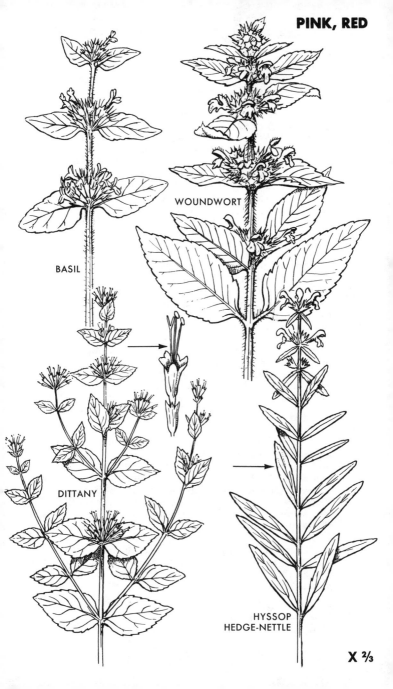

PINK, RED

WOUNDWORT

BASIL

DITTANY

HYSSOP
HEDGE-NETTLE

X ⅔

PEALIKE FLOWERS; VARIOUS LEAVES

REDBUD   **Flowerbuds, flowers, young pods**
Cercis canadensis
A small tree that becomes cloudy with clusters of small, pink or lavender pealike flowers before the leaves appear. Leaves broadly *heart-shaped* with pointed tips and smooth edges; 2–6 in. (5–15 cm) long. Fruit flattened pinkish pods. Usually small, but up to 50 ft. (15 m). **Where found:** Moist woods, thickets. Se. Neb., s. Wisc., Conn., south to Texas, n. Fla. **Flowers:** March–May. **Fruit:** July–Aug. (or longer).
Use: Pickle, salad, cooked vegetable. Redbud is one of the most conspicuous and attractive trees of early spring. The *flowerbuds* can be pickled, and the *flowers* added to salads. The buds, flowers, and tender *young pods* can be sautéed for 10 min. in butter.   EARLY SPRING (flowerbuds, flowers);
EARLY SUMMER (pods)

GROUNDNUT   **Tubers**
Apios americana
A small, twining vine with compact, fragrant clusters of *maroon or lilac-brown* flowers in the leaf axils. Leaves smooth, light green; with 5–7 ovate, sharp-pointed leaflets. Root a long string of walnut-sized tubers. **Where found:** Rich moist soil; thickets, streambanks, open woods. Minn. through Great Lakes area to New Brunswick, south to Texas and Fla. **Flowers:** June–Sept.
Use: Potato. Once they have been washed and peeled, the small tubers are excellent boiled (about 20 min.), roasted, or fried in bacon fat; the taste is sweet and mildly turniplike. Eat while still hot; they become much less palatable when cold.
ALL YEAR

BEACH PEA   **Seeds**
Lathyrus japonicus   **Color pl. 1**
Low, sprawling; with showy purple flowers on stiff-erect flowerstalks. Leaves terminated by curling tendrils; leaflets oval, paired or alternate, somewhat fleshy. Leafstalks embraced by *large arrowhead-shaped stipules*. Seedpods reminiscent of garden peas, but smaller. 1–2 ft. (30–60 cm). **Where found:** Sandy or gravelly shores, beaches. Coast south to N.J.; also shores of Great Lakes, Oneida Lake, Lake Champlain. **Flowers:** June–Aug.
Use: Cooked vegetable (peas). Gather the seeds while still small, tender, and bright green; older, they become rather dry and tasteless. Boil for 15–20 min. and serve with butter and pepper; eat while still hot. Rich in B vitamins, vitamin A, and protein. **Note:** See Beach Pea, p. 142.   SUMMER

GOAT'S-RUE   **Poisonous**
Tephrosia virginiana
Flowers bicolored (yellow and pink). See p. 82.

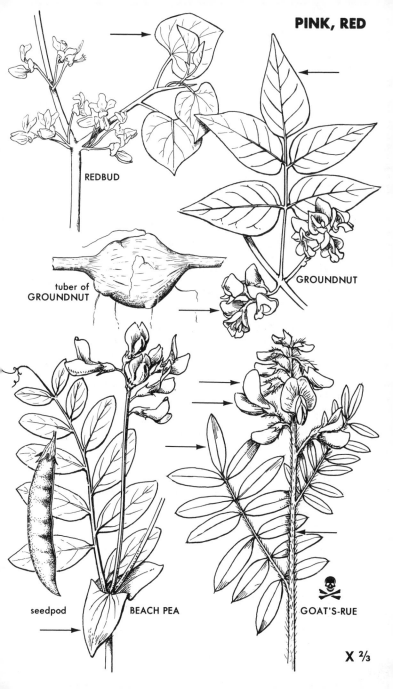

PINK, RED

REDBUD

GROUNDNUT

tuber of
GROUNDNUT

seedpod BEACH PEA

GOAT'S-RUE

X ⅔

PEALIKE FLOWERS; LEAFLETS IN 3'S

ALFALFA, LUCERNE　　　　　　**Young leaves and**
Medicago sativa　　　　　　　　**flowerheads**
Blue-violet, sometimes purple. See p. 142. SPRING–SUMMER

CLOVERS　　　　　　**Young leaves, flowerheads,**
Trifolium spp.　　　　　　　　　　　**seeds**
Familiar field and wayside weeds with compact flowerheads
and 3 leaflets. See also clovers, p. 56, and Hop Clover, p. 80.
Use: Salad, cooked green, flour, tea. Although not the choicest
of wild foods, clovers are both abundant and rich in protein.
The *flowerheads* and tender *young leaves* are difficult to digest
raw, but can be eaten in quantity if soaked for several hours in
salty water or boiled for 5–10 min. The dried *flowerheads* make
a delicate, healthful tea when mixed with other teas. The dried
flowerheads and *seeds* can be ground into a nutritious flour.
SPRING–SUMMER

CRIMSON or ITALIAN CLOVER, *T. incarnatum.* Flowerhead *longish,* blood-red. Leaflets *very blunt.* 6–30 in.
(15–75 cm). **Where found:** Cultivated. **Flowers:** May–July.
ALSIKE CLOVER, *T. hybridum.* Pale pink or whitish.
Leaves unmarked; *branching from stems.* 1–2 ft. (30–60 cm).
Where found: Fields, roadsides. Throughout. **Flowers:**
May–Oct.
RED CLOVER, *T. pratense.* Flowers purple-red, in rounded
heads. Leaves marked with *pale chevrons.* 6–16 in. (15–
40 cm). **Where found:** Fields, roadsides. Throughout. **Flowers:** April–Sept.

HOG-PEANUT　　　　　　　　**Subterranean seeds**
Amphicarpa bracteata
A low twining vine with slender stems and light green, ovate
leaflets. Small clusters of pale lilac to white, pealike flowers in
upper leaf axils produce curved pods with 3–4 mottled, inedible
seeds. Petal-less flowers on threadlike runners near base of
plant produce fleshy 1-seeded pods just below ground level.
Where found: Rich soil; moist woods, thickets. Most of our
area. **Flowers:** July–Sept.
Use: Cooked vegetable (beans). The light brown seeds from
the *subterranean* pods are somewhat dry but quite good boiled
for 15–20 min. and served with butter. FALL–EARLY SPRING

WILD BEAN　　　　　　　　　　　　　**Seeds**
Phaseolus polystachios
Similar to Hog-peanut (above), but coarser. Side leaflets more
asymmetrical in shape, with *deeply rounded bases.* Flowers
loosely arranged on a *long* stalk; red-purple. Pods curved, with
4–6 brown seeds. **Where found:** Dry woods, sandy thickets.
Iowa to s. N.J. and south. **Flowers:** July–Sept.
Use: Cooked vegetable (beans). Prepare the fresh or dried
seeds like cultivated beans. LATE SUMMER–EARLY FALL

124

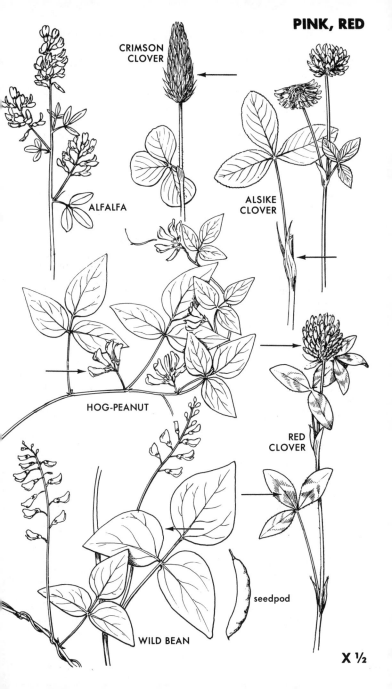

PINK, RED

CRIMSON CLOVER

ALFALFA

ALSIKE CLOVER

HOG-PEANUT

RED CLOVER

seedpod

WILD BEAN

X ½

THISTLELIKE OR DANDELIONLIKE FLOWERS

BURDOCKS   **Young leaves, roots; young leaf-**
Arctium spp.   **and flower-stalks (Great Burdock)**
Biennials producing large, bushy, rough, slightly woolly basal leaves
the first year and bushy flowerstalks with numerous purple-
flowered, thistlelike burs the second. 4 species; 2 shown.

Use: Cooked green, cooked vegetable, salad, candy. The tender
young leaves can be added to salads or boiled in several changes
of water. Once the thick inedible rind has been removed, the
roots of the first-year plants can be boiled for 30 min. in 2

changes of water and served with butter. The white pith (without the bitter green rind) from the *young leaf- and flowerstalks of Great Burdock* is excellent added to salads or prepared in the same way as the roots; once cooked, the flowerstalks can be simmered in a sugar syrup to make candy.

SPRING (leaves and leafstalks);
EARLY SUMMER (roots); SUMMER (flowerstalks)
GREAT BURDOCK, *A. lappa.* Burs large, up to 1½ in.
(3.8 cm), long-stalked; in flat-topped clusters. Stalks of lower
leaves solid with a groove on upper surface. 4–9 ft. (1.2–2.7 m).
Where found: Roadsides, waste ground, limy soil. E. Canada
to Ill., Penn., and New England. **Flowers:** July–Oct.
COMMON BURDOCK, *A. minus.* **Color pl. 9.** Burs much
smaller, *short-stalked.* Stalks of lower leaves *hollow,* not
grooved. 3–5 ft. (0.9–1.5 m). **Where found:** Roadsides, waste
ground. Canada south to Kans., Mo., W. Va., and Va. **Flowers:** July–Oct.

THISTLES   **Young leaves, young stems,**
Cirsium spp.   **roots**
Biennials with deeply-cut, prickly leaves and showy, rose-pur-
ple, "shaving-brush" flowers. Numerous species; 1 shown.

Use: Salad, cooked green, cooked vegetable. With the spines
removed, the *young leaves* can be added to salads or cooked as
greens. The pithy *young stems* are excellent peeled and eaten
raw or cooked. The raw or cooked *roots* of first-year plants
(those without stems) are a fine survival food. SPRING–FALL
BULL THISTLE, *C. vulgare.* **Color pl. 10.** Note the
prickly-winged stems and the *rigid, yellow-tipped spines* on the
bracts of the 1–3 large flowerheads. 2–6 ft. (0.6–1.8 m). **Where
found:** Pastures, roadsides. Throughout. **Flowers:** June–Oct.

OYSTER-PLANT, SALSIFY   **Young leaves, roots**
Tragopogon porrifolius
Similar to Yellow Goat's-beard but with *purple* rays. 2–5 ft.
(0.6–1.5 m). **Where found:** Roadsides, fields. Ontario to Nova
Scotia, south to Kans., Ga. **Flowers:** May–July.
Use: Salad, cooked green, cooked vegetable, coffee. See Yellow
Goat's-beard (*T. pratensis*), p. 84.   SPRING (leaves);
FALL–EARLY SPRING (roots)

126

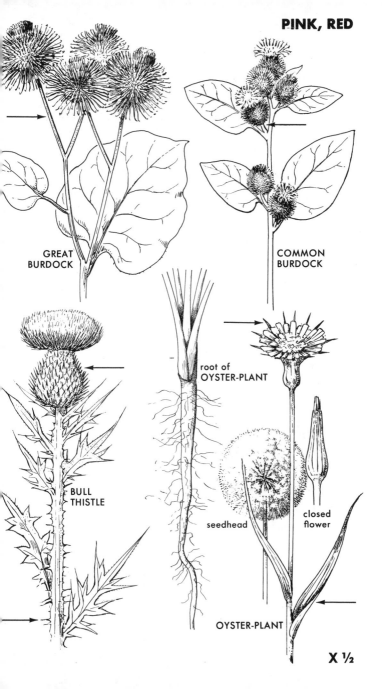

GREAT
BURDOCK

COMMON
BURDOCK

BULL
THISTLE

root of
OYSTER-PLANT

seedhead

closed
flower

OYSTER-PLANT

X ½

MISCELLANEOUS BLUE
OR VIOLET FLOWERS

VIRGINIA WATERLEAF   **Young leaves**
Hydrophyllum virginianum
Note the radiating clusters of white or pale violet, 5-petaled
flowers with conspicuously protruding stamens and irregularly
cut *5- to 7-lobed leaves* (often *marked* as if stained with water).
Plant smoothish; basal leaves in rosettes. 1–3 ft. (30–90 cm).
See also p. 44. **Where found:** Rich woods, damp clearings.
Manitoba to Quebec, south to e. Kans., n. Ark., Tenn., Va., and
w. New England. **Flowers:** April–Aug.
Use: Cooked green. The young leaves, before the flowers appear, are excellent boiled for 5–10 min. in 1 or 2 changes of
water and served with vinegar.   EARLY SPRING

LESSER BROOMRAPE   **Stem**
Orobanche minor
Note the scaly, leafless, unbranched, yellow-brown, downy stem
(much of which is underground) and the *stout spike* of violet
flowers. 4–18 in. (10–45 cm). **Where found:** Parasitic on roots
of clover or tobacco. N.J. to N.C. **Flowers:** April–July.
Use: Asparagus. Several authors suggest that the tender underground portions of the stems can be prepared like asparagus. A similar species was used by the Paiute Indians in the
West.   EARLY SUMMER (usu. June)

CREEPING BELLFLOWER   **Roots**
Campanula rapunculoides
Delicate 5-pointed violet-blue bells droop along 1 side of a stiff
stem; flowers bloom from the bottom up. Plants spread by
creeping runners. 1–3 ft. (30–90 cm). **Where found:** A garden
escape; fields, roadsides. S. Canada south to Mo., Ill., Ind.,
Ohio, W. Va., Md., Del. **Flowers:** June–Sept.
Use: Salad, cooked vegetable. The slender runners send down
fleshy underground branches which can be chopped and added
to salads, or boiled for 20 min. The taste is slightly sweet,
suggesting parsnips.   LATE SUMMER–FALL

AMERICAN BROOKLIME ·   **Leaves and stems**
Veronica americana
A smooth, *succulent* plant with prostrate to erect stems and
toothed, ovate or long-ovate, *stalked* leaves. Small 4-petaled
blue flowers (inset), typical of speedwells, *Veronica* spp., are at
the top of long stems originating in the leaf axils. 3–10 in.
(7.5–25 cm). **Where found:** Shallow water, springs, brooks,
swamps. Canada south to Neb., N.C. **Flowers:** May–Aug.
Use: Salad. The succulent leading tips can be used like watercress. Although somewhat bitter, many find the mildly pungent taste excellent.   LATE SPRING–SUMMER

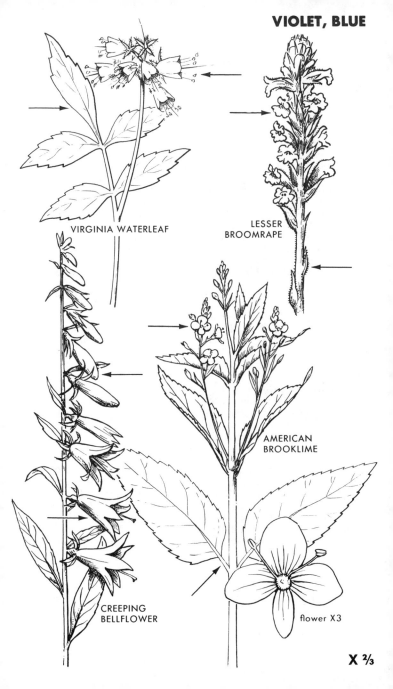

VIOLET, BLUE

VIRGINIA WATERLEAF

LESSER
BROOMRAPE

AMERICAN
BROOKLIME

CREEPING
BELLFLOWER

flower X3

X ⅔

PETALS IN 3'S:
IRIS, SPIDERWORTS

 DAYFLOWERS   **Young leaves and stems**
Commelina spp.

Erect or reclining 3-petaled plants. The lower petal is smaller than those above and supports curved stamens. Note the heart-shaped spathe just below the flowers. Several species; 2 shown.

Use: Salad, cooked green. The young leaves and stems can be added fresh to salads, or boiled for 10 min. and served with butter.   SPRING–EARLY SUMMER

VIRGINIA DAYFLOWER, *C. virginica.* Erect. Lower petal *blue.* 1½–3 ft. (45–90 cm). **Where found:** Rich low woods, thickets, clearings. Kans., Mo., s. Ill., Ky., W. Va., Md., s. Penn., and s. N.J., south to Texas and n. Fla. **Flowers:** July–Oct.

ASIATIC DAYFLOWER, *C. communis.* Reclining. Lower petal smaller than in preceding, *white.* Leaves wider. **Where found:** Roadsides, waste places, edges. Wisc. to Mass., south to e. Kans., Ark., Miss., Ala., and Ga. **Flowers:** May–Oct.

BLUE FLAG, WILD IRIS   **Poisonous**
Iris spp.

Familiar plants with showy violet (also red or yellow) flowers. LARGER BLUE FLAG, *I. versicolor,* is closely similar to garden iris, with blue-green swordlike leaves and prominently veined, downcurved violet sepals. 2–3 ft. (60–90 cm). **Where found:** Marshes, wet meadows. Manitoba to s. Labrador, south to Minn., Wisc., Ohio, Va. **Flowers:** May–July.

Warning: Do not confuse the rootstocks of the irises with those of Sweetflag (p. 62) or Cattail (p. 158); all irises are poisonous. Iris rootstocks are odorless and very unpleasant tasting; Sweetflag rootstocks are pleasantly aromatic, and Cattail rootstocks are odorless and bland tasting.

SPIDERWORT   **Young leaves and stems,**
Tradescantia virginiana   **Color pl. 10**   **flowers**

 Showy clusters of 3-petaled violet flowers terminate a succulent stem. Petals roundish, symmetric; stamens golden. Leaves also succulent; long, irislike. A variety of *Tradescantia* species in our area, particularly southward; all are edible, but this species considered the best. 1–2½ ft. (30–75 cm). **Where found:** Wood edges, thickets, meadows, roadsides. Wisc. to Conn., south to Mo., Tenn., nw. Ga.; escaped from cultivation in ne. U.S. to Maine. **Flowers:** April–July.

Use: Salad, cooked green, candy. The young *leaves* and *stems* can be added fresh to salads or boiled for 10 min. The *flowers* can be candied: Taking care not to bruise the petals, dip the flowers in water, dry, carefully brush with slightly beaten egg white, cover with sugar, and dry.   SPRING (leaves);
LATE SPRING–EARLY SUMMER (flowers)

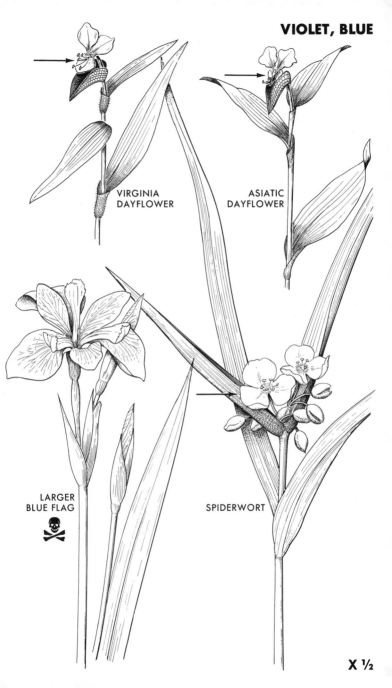

VIOLET, BLUE

VIRGINIA
DAYFLOWER

ASIATIC
DAYFLOWER

LARGER
BLUE FLAG

SPIDERWORT

X ½

SPURRED OR HOODED FLOWERS

VIOLETS **Young leaves, flowers**
Viola spp.
Familiar spring flowers; blue-violet (also yellow or white), 5-petaled, with lowest petal heavily veined and extending back into a spur, and lateral petals usually bearded. Most violets are edible, but some yellow species may be mildly cathartic.

Use: Salad, cooked green, soup thickener, tea, candy. The tender young *leaves* can be added to salads, boiled for 10-15 min. to make a palatable cooked green, or added to soups as an okralike thickener; violet leaves are somewhat bland and are best mixed with other greens. The dried leaves can be made into tea. The *flowers* can be candied (see Spiderwort, p. 130). Leaves rich in vitamins A and C.   EARLY SPRING–SPRING
MARSH BLUE VIOLET, *V. cucullata.* Flowerstalk *rises above* heart-shaped leaves. Lower petal *shorter* than in following; side petals darker toward throat. 5-10 in. (12.5-25 cm). **Where found:** Wet meadows, springs, bogs. S. Canada south to Neb., Tenn., Ga. (mts.). **Flowers:** April–June.
COMMON BLUE VIOLET, *V. papilionacea.* **Color pl. 2.** Flowers *barely exceed* heart-shaped leaves. 2 side petals bearded; lower petal longer, unbearded; all 3 lower petals boldly veined. 3-8 in. (7.5-20 cm). **Where found:** Damp woods, meadows. N. Dak. to cen. Maine, south to Okla., Ga. **Flowers:** March–June.
BIRDFOOT VIOLET, *V. pedata.* Note the deeply cut leaves. Flowers either bicolored (3 lower petals paler), or uniformly colored. Upper petals curve backwards; all petals *beardless.* Orange tips of stamens conspicuous at center. 4-10 in. (10-25 cm). **Where found:** Sandy fields, slopes, sunny rocks. Much of our area. **Flowers:** April–June.

DWARF or SPRING LARKSPUR **Poisonous**
Delphinium tricorne
Note the loose cluster of *erectly spurred* blue or white flowers and 5- or 7-lobed buttercuplike leaves. 12-30 in. (30-75 cm). **Where found:** Rich woods, slopes. Minn. to Penn., south in uplands to Okla., Ark., Ala., and Ga. **Flowers:** April-May.
Warning: The young foliage and seeds of any of the larkspurs can be fatal if eaten in large quantities.

MONKSHOOD **Poisonous**
Aconitum uncinatum
Note the large *helmetlike* upper sepal (covering 2 petals) and buttercuplike leaves. 2-3 ft. (60-90 cm). **Where found:** Low woods, damp slopes. Ind. to Penn., south in mts. to Ala. and Ga. **Flowers:** Aug.-Oct.
Warning: All parts of this plant are highly poisonous. Even a nibble can cause harm.

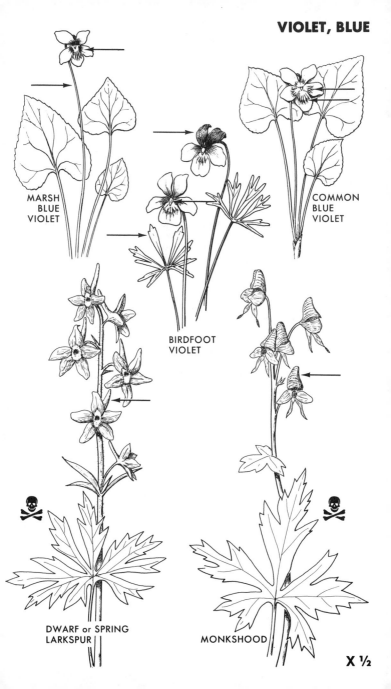

VIOLET, BLUE

MARSH BLUE VIOLET

COMMON BLUE VIOLET

BIRDFOOT VIOLET

DWARF or SPRING LARKSPUR

MONKSHOOD

X ½

5-PETALED FLOWERS:
MISCELLANEOUS

 COMFREY   **Young leaves**
Symphytum officinale
White, cream, pink, purplish, or pale blue. See p. 110.
                                        EARLY SPRING–SPRING

CORN-SALAD   **Young leaves**
Valerianella olitoria
Delicate; with forking stems and opposite, oblong, stalkless
leaves. Flowers in *small flat clusters surrounded by leafy
bracts;* tiny, pale blue. Low rosettes of thin, pale green basal
leaves appear in the fall; the stems and flowers the next spring.
4–12 in. (10–30 cm). (Several taller species with white flowers
also edible.) **Where found:** Waste ground. Ind. to New England, south to Tenn., N.C. **Flowers:** April–May.
Use: Salad, cooked green. The tender young leaves are excellent added to salads or prepared like spinach; gather them
before the flowers appear.   FALL–EARLY SPRING

 CHEESES, COMMON MALLOW   **Young leaves, green**
Malva neglecta   **Color pl. 8**   **fruit**
Pale blue-lavender, lilac-pink, or white. See p. 108.
                    SPRING (leaves)–SUMMER (leaves, fruit)

VIOLET WOOD-SORREL   **Leaves**
Oxalis violacea
Purple-violet to rose-purple. See p. 104.   SPRING–SUMMER

 BLUE VERVAIN   **Seeds**
Verbena hastata
Note the clusters of tiny, intensely blue flowers that advance
toward the tips of the slender, branching flower spikes. Leaves
opposite, *narrow,* toothed; lower leaves occasionally 3-lobed.
Stem grooved, 4-sided. **Where found:** Damp thickets, roadsides, wet meadows. Most of our area. **Flowers:** June–Oct.
Use: Flour. Roasted and ground, the tiny seeds make a mildly
bitter but palatable flour. Soaking the seeds in several changes
of cold water reduces the bitterness.   LATE SUMMER–FALL

NIGHTSHADE, BITTERSWEET   **Poisonous**
Solanum dulcamara
A weak woody-stemmed vine trailing over bushes. Note the
recurved violet petals and *beaklike* yellow anthers. Leaves
usually with *2 small lobes* at bases. Berries oval, in drooping
clusters; green, then later bright red. 2–8 ft. (0.6–2.4 m).
Where found: Moist thickets, clearings. Minn., Ontario, Nova
Scotia south to Kans., Mo., and Ga. **Flowers:** May–Oct.
Warning: The bitter ruby-red berries are quite visible and
attractive looking, but very poisonous.

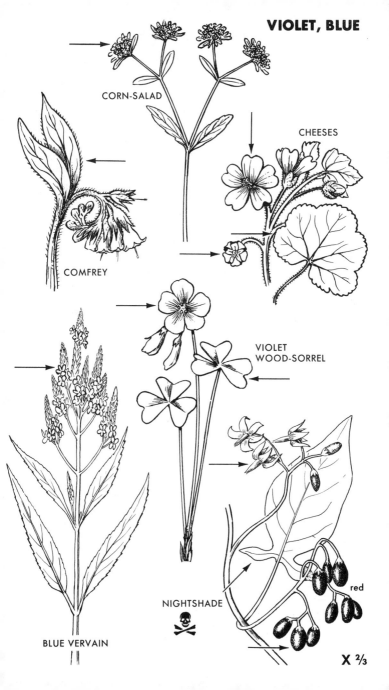

VIOLET, BLUE

CORN-SALAD

CHEESES

COMFREY

VIOLET
WOOD-SORREL

BLUE VERVAIN

NIGHTSHADE
☠

red

X ⅔

6 PETALS; SHOWY SPIKES

PICKERELWEED   **Young leaves, fruit (seeds)**
Pontederia cordata   **Color pl. 3**
Widespread aquatic plants forming beds in shallow water.
Note the conspicuous blue spike atop a thick fleshy stem and
the large *arrowhead-shaped* leaves. 1–4 ft. (0.3–1.2 m). **Where
found:** Pond edges, shallow water. Minn., s. Ontario, Nova
Scotia, south through most of our area. **Flowers:** May–Oct.
Use: Salad, cooked green, cereal, flour. Before fully unfurled,
the young *leaves* can either be chopped and added to salads, or
boiled for 10 min. and served with butter. The distinctive *fruit*
(inset), each containing a single starchy seed, are highly nutritious and can be eaten out of hand, dried and added to granola-
like cereals, or roasted and ground into flour.

        EARLY SUMMER (leaves); LATE SUMMER–FALL (fruit)

WATER-HYACINTH   **Young leaves and leafstalks,**
Eichhornia crassipes   **flowers**
An abundant *floating* plant that often chokes southern water-
ways. Showy 6-lobed bluish-purple flowers cluster on a spike in
the center of a floating rosette of leaves; upper lobes of flowers
have yellow centers. Leaf blades roundish, 2–5 in. (5–12.5 cm)
wide; leafstalks cylindrical, swelling balloonlike at bases. 5–30
in. (12.5–75 cm). **Where found:** Quiet or slow-moving water.
Mo. to Va., and south to e. Texas and Fla.; commonest in deep
South. **Flowers:** June–Aug. (earlier and later in deep South).
Use: Cooked vegetable. The *young leaves and leafstalks,* while
still light green, are surprisingly good either boiled thoroughly
or fried until crisp. The *flowers* can also be boiled, becoming
somewhat gelatinous. **Warning:** Water-hyacinth may cause
an itching sensation if eaten raw; this property is usually removed by cooking, but not necessarily. Try sparingly at first.
                                      SPRING–SUMMER

WILD HYACINTH, EASTERN CAMASS   **Bulb**
Camassia scilloides
Pale blue *6-petaled* stars top a simple stem that rises above a
basal rosette of *keeled* linear leaves. Root an onionlike bulb.
1–2 ft. (30–60 cm). **Where found:** Meadows, moist woods.
Iowa, s. Wisc., se. Mich., s. Ontario, sw. Penn., south to Texas
and Ala.; mostly west of Appalachians. **Flowers:** April–June.
Use: Cooked vegetable. The onionlike bulbs can either be
boiled for 20–30 min., or baked in aluminum foil for 45 min. at
350°F (173°C); the results are pleasant tasting, but gummy
textured. Although edible all year, the bulbs are most easily
located and identified when the plants are in bloom. Rich in
sugar.   SPRING

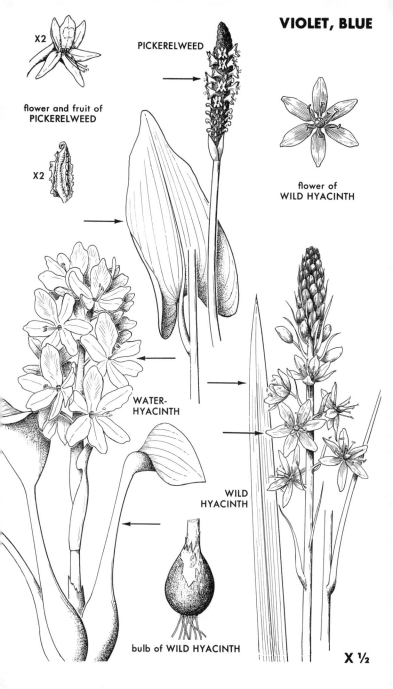

VIOLET, BLUE

X2

flower and fruit of
PICKERELWEED

X2

PICKERELWEED

flower of
WILD HYACINTH

WATER-
HYACINTH

WILD
HYACINTH

bulb of WILD HYACINTH

X ½

SQUARE STEMS; PAIRED LEAVES:
MINTS (1)

MINTS **Leaves**
Mentha spp.
Familiar aromatic plants with square stems, paired leaves, and small lipped flowers clustered in leaf axils or terminal spikes. A variety of aromatic mints throughout our area; all are potential teas; 4 species shown.
Use: Tea, flavoring. The fresh or dried leaves make an excellent tea when steeped in hot water for 5–10 min. They can also be used to flavor jellies, sauces, dressings, drinks. SUMMER

PEPPERMINT, *M. piperita.* **Color pl. 6.** Note the familiar odor and *hot taste* of the smooth, short-stalked leaves. Stems branching, purplish. Flowers pink to pale violet, on a short *or interrupted spike* (see also p. 118). $1\frac{1}{2}$–3 ft. (45–90 cm). **Where found:** Wet meadows, streambanks. Throughout. **Flowers:** June–Oct.

WATER MINT, *M. aquatica.* Similar to Peppermint, but leaves *round-ovate.* Flowerheads wider, pale lavender. Stems *and* leaves often purplish. 18–30 in. (45–75 cm). **Where found:** Wet places. Nova Scotia to Del. **Flowers:** Aug.–Oct.

SPEARMINT, *M. spicata.* Note the characteristic odor of the virtually *stalkless,* smooth leaves. Flowers pink to pale violet, in slender interrupted clusters; stamens protruding. 10–20 in. (25–50 cm). **Where found:** Wet places. Throughout. **Flowers:** June–Oct.

WILD MINT, *M. arvensis.* **Color pl. 2.** Note the strong minty odor, and downy or hairy unbranched stems. Note also the tight clusters of tiny bell-shaped flowers in the leaf axils; pale violet, lavender, or white (see p. 54). 6–24 in. (15–60 cm). **Where found:** Damp soil, shores. Canada, n. U.S. **Flowers:** July–Sept.

EUROPEAN HORSEMINT, *M. longifolia.* Coarser than Spearmint (above), but with longer, milder-smelling leaves. Stems *downy-hairy;* leaves woolly-white beneath. Flowers in slender interrupted clusters. 1–$2\frac{1}{2}$ ft. (30–75 cm). **Where found:** Roadsides, thickets. Ind. to Mass., south to Mo., Va. **Flowers:** July–Sept.

DOWNY WOOD-MINT **Leaves**
Blephilia ciliata
Note the whorls of pale bluish-purple flowers separated by a row of colored bracts. Stem leaves almost without stalks, white-downy beneath. 1–3 ft. (30–90 cm). HAIRY WOOD-MINT, *B. hirsuta* (not shown), is hairy, and has long-stalked leaves and looser flower clusters. **Where found:** Dry woods, thickets. Wisc. to Vt., south to e. Texas, Okla., Ark., Miss., and Ga. **Flowers:** May–Aug.
Use: Tea. The mildly aromatic leaves can be prepared like those of the true mints (above). SUMMER

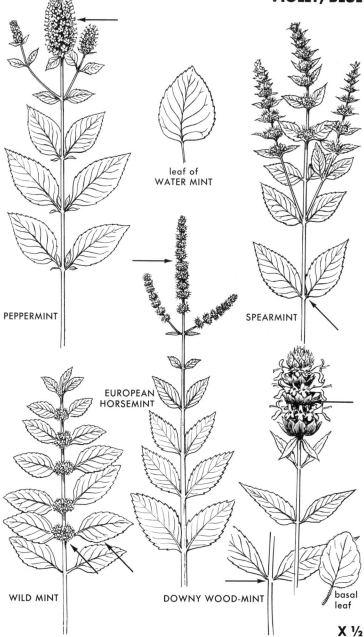

VIOLET, BLUE

leaf of
WATER MINT

PEPPERMINT

SPEARMINT

EUROPEAN
HORSEMINT

WILD MINT

DOWNY WOOD-MINT

basal
leaf

X ½

SQUARE STEMS; PAIRED LEAVES:
MINTS (2)

 GILL-OVER-THE-GROUND, GROUND-IVY   **Leaves**
Glechoma hederacea   **Color pl. 8**
A creeping or trailing weed of lawns and shady places. Leaves round or kidney-shaped, scalloped-toothed; often tinged with purple. Flowers blue-lavender, loosely clustered in leaf axils. **Where found:** Lawns, roadsides, shady spots. Se. Canada south to Kans., Mo., Ala., Ga. **Flowers:** March–July.
Use: Tea. The dried leaves make a fine herbal tea when steeped for 5–10 min. in hot water.   SPRING–SUMMER

 AMERICAN PENNYROYAL   **Leaves**
Hedeoma pulegioides
A small, usually inconspicuous mint with *highly aromatic* leaves. Soft-hairy, sometimes branching. Flowers small, pale violet or bluish; clustered in leaf axils. Calyx characteristically with 3 short, broad teeth above and 2 long, narrow, curved teeth below. 6–18 in. (15–45 cm). **Where found:** Dry soil, fields, roadsides, open woods. Minn. to Quebec, south to Okla., Ark., Ala., and Fla. **Flowers:** July–Sept.
Use: Tea. The fresh or dried leaves smell strongly of mint; they make an excellent tea when steeped in hot water for 5–10 min. Dried leaves should be stored in an airtight container.   SUMMER

 BLUE GIANT HYSSOP   **Leaves**
Agastache foeniculum
Note the smooth square stem and paired leaves that are white-downy beneath and *smell of anise.* Flowers blue-violet, on interrupted spikes at end of stem and in leaf axils. Each flower (inset) with 2 pairs of long protruding stamens; 1 pair curves upward, the other downward, *crossing.* 2–4 ft. (0.6–1.2 m). **Where found:** Dry ground, thickets, prairies. Prairie states; Manitoba to Ontario, south Iowa, Ill.; locally eastward. **Flowers:** June–Sept.
Use: Tea. The fresh or dried leaves make a delicate anise-flavored tea when steeped for about 10 min. in hot water.
    SUMMER

 CATNIP   **Leaves**
Nepeta cataria
The much-branched leafy stems are terminated by dense clusters of pale violet or white flowers with purple spots. Leaves jagged, arrowhead-shaped; gray-green with soft whitish down. Minty odor mildly unattractive. 6–24 in. (15–60 cm). **Where found:** Waste ground, roadsides. Throughout. **Flowers:** June–Sept.
Use: Tea. The dried leaves make a pleasantly soothing tea when prepared like oriental tea.   SUMMER

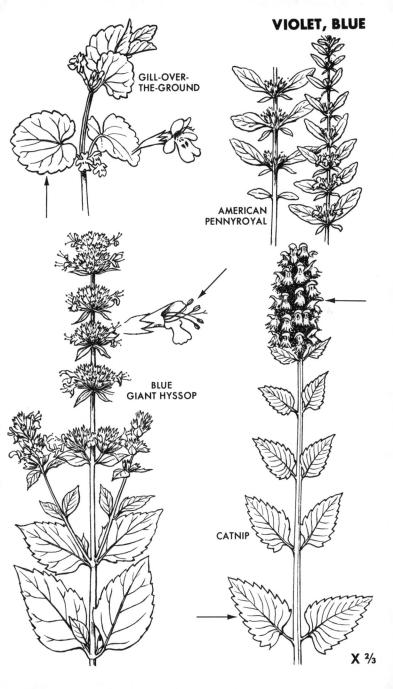

VIOLET, BLUE

GILL-OVER-THE-GROUND

AMERICAN PENNYROYAL

BLUE GIANT HYSSOP

CATNIP

X ⅔

PEALIKE FLOWERS

BEACH PEA   **Seeds**
Lathyrus japonicus   **Color pl. 1**
A low, somewhat fleshy plant with showy pink-lavender to violet, pealike flowers. Note the broad *arrowhead-shaped stipules* clasping the leafstalks. Leaves terminated by curling tendrils; leaflets oval, paired or alternate. Seedpods reminiscent of garden peas, but smaller. 1–2 ft. (30–60 cm). **Where found:** Sandy shores, beaches. Coast south to N.J.; also shores of Great Lakes, Oneida Lake, Lake Champlain. **Flowers:** June–Aug.
Use: Cooked vegetable (peas). Gather the seeds while still small, tender, and bright green; as they get older, the seeds become rather dry and tasteless. Boil for 15–20 min. and serve with butter and pepper. Be sure to eat the peas while still hot. Rich in B vitamins, vitamin A, and protein. **Note:** Although this species is perfectly edible, other members of the *Lathyrus* genus are poisonous.   SUMMER

ALFALFA, LUCERNE   **Young leaves and flowerheads**
Medicago sativa
A bushy or prostrate weed with cloverlike leaves and small — less than ½ in. (1.3 cm) long — blue to violet (sometimes purple), pealike flowers in short spikes. Seedpods twist spirally. 1–1½ ft. (30–45 cm). **Where found:** Fields. Cultivated, but often a roadside escape. Throughout; least common in Southeast. **Flowers:** May–Oct.
Use: Tea, nutritional supplement, salad. The dried and powdered young leaves and flowerheads are highly nutritious. They can be steeped in hot water for 10 min. to make a healthful tea, or added to breakfast cereals or stews as a nutritional supplement. The taste is fairly bland, so when making tea, it is best to mix Alfalfa with other teas. The tender young leaves can be added fresh to salads. Rich in vitamins A, D, and K.   SPRING–SUMMER

WILD LUPINE   **Poisonous**
Lupinus perennis
Note the showy spikes of blue or violet (rarely pink or white) pealike flowers and *palm-shaped* leaves (radiating into 7–9 segments). 1–2 ft. (30–60 cm). **Where found:** Dry open woods, clearings. Minn., s. Ontario, N.Y., s. Maine, south to La. and Fla. The larger garden lupines (not shown) are roadside escapes in New England and mts. **Flowers:** April–July.
Warning: The pealike seeds have been wrongly recommended by some authors as a possible substitute for peas; they contain a poisonous alkaloid and should be avoided.

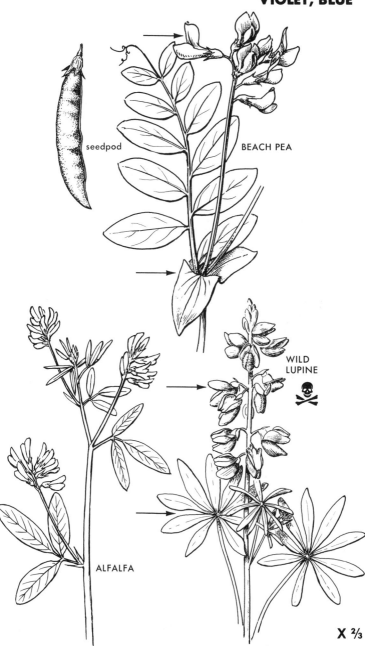

VIOLET, BLUE

seedpod

BEACH PEA

WILD
LUPINE

ALFALFA

X ⅔

DAISYLIKE OR
DANDELIONLIKE FLOWERS

CHICORY　　　　　　　　　　**Young leaves, roots**
Cichorium intybus　　**Color pl. 7**
Note the stiff, nearly naked stems with their strikingly blue,
stalkless flowers — 1½ in. (3.8 cm). Flower rays blue (often
white, rarely pink), square-tipped and fringed; closed by after-
noon or when overcast. Basal leaves dandelionlike; sap milky.
Root a fleshy white taproot. To 4 ft. (1.2 m). **Where found:**
Roadsides, waste places. Widespread. **Flowers:** May–Oct.
Use: Coffee, salad, cooked green. The *roots* make an excellent
coffeelike beverage when roasted in an oven until dark brown
and brittle, ground, and prepared like coffee; use roughly 1½
tsp. Chicory for each cup of water. The white underground
parts of the young *leaves* are a welcome addition to salads, and
the aboveground parts are excellent boiled for 5–10 min. in just
enough water to cover. Gather the leaves early; they soon
become too bitter to use.　　　　EARLY SPRING (leaves);
　　　　　　　　　　　　　FALL–EARLY SPRING (roots)

LARGE-LEAVED ASTER　　　　**Very young leaves**
Aster macrophyllus
Note the *sticky-branched,* slightly flat-topped clusters of
many-rayed violet flowers; central disks of flowers turn reddish. Basal leaves 4–8 in. (10–20 cm) wide, heart-shaped, rough;
usually with a broad basal notch. Upper leaves small, stalkless. Stems purplish. 1–5 ft. (0.3–1.5 m). **Where found:**
Woods, clearings. Minn. to maritime Canada, south to Ill.,
Md.; in mts. to N.C. **Flowers:** Late July–Sept.
Use: Cooked green. Several authors suggest this as a possible
spinachlike cooked green. Only the tenderest, youngest leaves
should be used as they soon become tough and leathery.
　　　　　　　　　　　　　　　　　　　SPRING

BLUE LETTUCES　　　**Young leaves, developing**
Lactuca spp.　　　　　　　　　　**flowerheads**
Tall, leafy plants with loose clusters of small blue or white
dandelionlike flowers. Sap milky, bitter. 2 species shown.
Use: Salad, cooked green. See wild lettuces, p. 86.
　　　SPRING (leaves); SUMMER (developing flowerheads)
BLUE LETTUCE, *L. floridana*. Flowers small, ½ in.
(1.3 cm), numerous; sometimes white. Leaves normally deeply
lobed, dandelionlike, but there is an unlobed form (var.
villosa). 3–7 ft. (0.9–2.1 m). **Where found:** Moist thickets,
woods. Minn. to se. Mass. and south. **Flowers:** Aug.–Oct.
BLUE LETTUCE, *L. pulchella*. More western. Pale, with
larger, less numerous flowers. Leaves narrowly lobed, or not at
all. **Where found:** Prairies, riverbanks. W. U.S. east to Mich.,
Mo., Okla. **Flowers:** July–Sept.

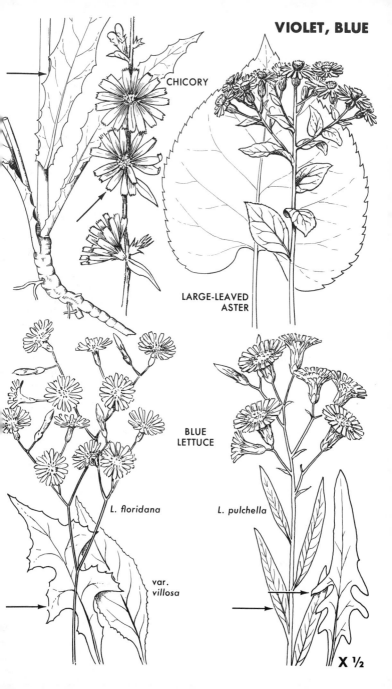

VIOLET, BLUE

CHICORY

LARGE-LEAVED ASTER

BLUE LETTUCE

L. floridana

var. *villosa*

L. pulchella

X ½

MISCELLANEOUS GREEN FLOWERS

GLASSWORTS   **Tender stems**
Salicornia spp.
Leafless plants with *succulent, jointed* stems that are light
emerald-green in summer and often red in fall. Erect or
sprawling, branched or sparingly so. Flowers tiny, hidden in
joints. SLENDER GLASSWORT, *S. europaea,* with joints
longer than broad, shown. 2–16 in. (5–40 cm). **Where found:**
Coastal saltmarshes, alkaline soil inland (widely scattered).
Nova Scotia south. **Flowers:** Aug.–Oct.
Use: Salad, purée, pickle. The tender stems can be added to
salads, puréed, or pickled. Salty; excellent.   SPRING–FALL

TOADSHADE, SESSILE TRILLIUM   **Young**
Trillium sessile   **unfolding leaves**
Maroon or yellowish green. See p. 96.   EARLY SPRING

SEA-BLITE   **Tender leaves and stems**
Suaeda maritima
A tiny, often prostrate, seaside plant. Leaves narrow, pointed,
fleshy, with a whitish bloom. Flowers tiny, buttonlike, tucked
singly in leaf axils. 3–12 in. (7.5–30 cm). **Where found:** Saltmarshes, beaches. Quebec to Va. TALL SEA-BLITE,
S. linearis (not shown), found along the coast from Maine to
Fla. and Texas, is similar but taller. **Flowers:** July–Oct.
Use: Cooked green. The tender leaves and stems are boiled for
10–12 min. in 2 or 3 changes of water; the changes of water
reduce their saltiness. An excellent source of salt for soups and
stews; not much is needed.   SPRING–LATE SUMMER

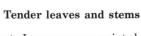

GINSENG   **Roots, leaves**
Panax quinquifolius
Note the 3 long-stalked, palm-shaped leaves (each with 5 irregularly toothed, stalked leaflets) in a whorl at the end of the
short green stem. Flowers pale yellow-green; in a small,
rounded, slender-stalked cluster. Root parsniplike, often
forked, aromatic. Berries red. 8–16 in. (20–40 cm). **Where
found:** Cool, rich woods. Manitoba to Quebec, south to Okla.,
La., Ala., and n. Fla. **Flowers:** Late May–July.
Use: Emergency food, tea. The parsniplike *roots* can be used
as an emergency food either raw or cooked. In addition, the
leaves can be made into tea. **Note:** Over-collection for sale as a
medicinal herb has made this once common plant quite rare; it
should not be used unless absolutely necessary.   SUMMER

SWAMP SAXIFRAGE   **Young unrolling leaves**
Saxifraga pensylvanica
Whitish, greenish, yellowish, purplish. See p. 48.   SPRING

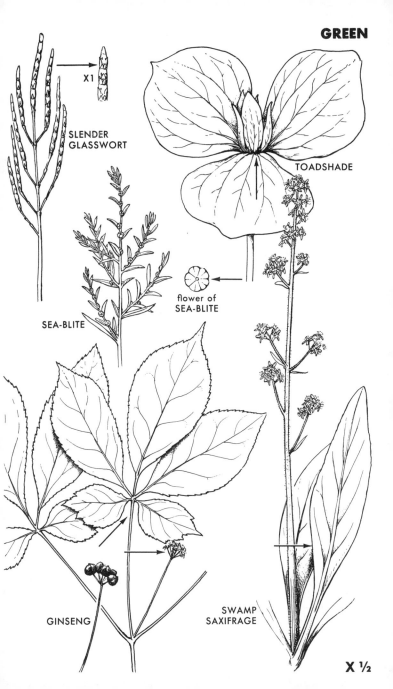

GREEN

SLENDER
GLASSWORT

X1

TOADSHADE

flower of
SEA-BLITE

SEA-BLITE

GINSENG

SWAMP
SAXIFRAGE

X ½

6-PART FLOWERS; PARALLEL VEINS

SOLOMON'S-SEAL **Young shoots, rootstock**
Polygonatum biflorum
Yellow or yellow-green. See p. 76. EARLY SPRING (shoots);
ALL YEAR (rootstock)

GREENBRIERS, CATBRIERS **Young shoots and**
Smilax spp. **leaves, rootstock**
Green-stemmed; usually woody or semi-woody. See p. 196.
LATE SPRING–EARLY SUMMER (shoots, leaves);
ALL YEAR (rootstock)

CARRION-FLOWER **Young shoots, rootstock**
Smilax herbacea **Color pl. 12**
A climbing vine similar to the greenbriers (above) but it *lacks thorns* and is not woody. Flowers smell of carrion. Berries blue-black, in globular clusters at the end of a long stem. **Where found:** Thickets, woods. S. Canada south to Mo., Tenn., and Ga. **Flowers:** April–June. **Fruit:** Aug.–Oct.
Use: Asparagus, jelly, flour. The tender *young shoots* are delicious boiled for 15 min. and served like asparagus. The *rootstocks* can be used like those of greenbriers, p. 196.
EARLY SPRING (shoots); ALL YEAR (rootstock)

CLINTONIA, CORN-LILY **Young leaves**
Clintonia borealis **Color pl. 15**
Yellow or yellowish green. See p. 74. EARLY SPRING

FALSE HELLEBORE, INDIAN POKE **Poisonous**
Veratrum viride
Large, stout, leafy-stemmed; topped by a large cluster of star-shaped yellowish green flowers. Leaves conspicuous in early spring; large, clasping, *heavily ribbed,* pleated. 2–8 ft. (0.6–2.4 m). **Where found:** Wet woods, swamps. S. Canada south to Md.; in uplands to Ga. **Flowers:** May–July.
Warning: Do not mistake for Skunk Cabbage, p. 156; here, leaf-veins parallel, and skunk-like odor lacking.

TWISTED-STALK **Young shoots, fruit**
Streptopus amplexifolius
Stems light-green, forking, zigzag; leaves alternate, clasping. A greenish white flower with reflexed petals hangs on a *crooked,* threadlike stalk below each leaf. Fruit an oval red berry. 1–3 ft. (30–90 cm). **Where found:** Cool woods, thickets. Canada, n. U.S.; in mts. to N.C. **Flowers:** May–July.
Use: Salad, nibble. The *young shoots* are an excellent cucumber-like addition to salads. Although mildly cathartic, a few *berries* make a pleasant nibble. **Note:** ROSE TWISTED-STALK, *S. roseus* (not shown), with pink blossoms, edible.
EARLY SPRING (shoots);
LATE SUMMER (fruit)

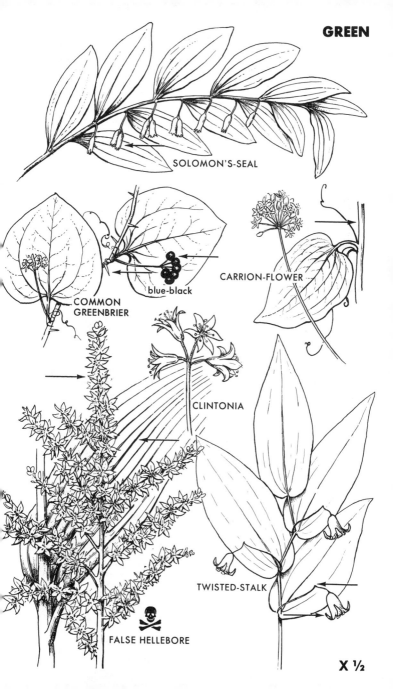

GREEN

SOLOMON'S-SEAL

COMMON GREENBRIER

blue-black

CARRION-FLOWER

CLINTONIA

FALSE HELLEBORE

TWISTED-STALK

X ½

SLENDER CLUSTERS IN LEAF AXILS; TOOTHED LEAVES; STINGING HAIRS

WOOD-NETTLE   **Young shoots and leaves**
Laportea canadensis
Somewhat similar to Stinging Nettle (below) but with fewer stinging hairs and with flower clusters terminal as well as in axils of upper leaves. Leaves long-stalked, *alternate;* ovate, with bases rounded or wedge-shaped rather than heart-shaped. Often in dense stands. 1–3½ ft. (0.3–1.1 m). **Where found:** Rich moist soil; woods, streambanks. Canada south to Okla., Miss., Ala., and Fla. **Flowers:** June–Sept.
Use: Cooked green, soup, tea. See nettles (below). Many consider this plant to be the best of the nettles.
SPRING (shoots)–SUMMER (tender leaves)

NETTLES   **Young shoots and leaves**
Urtica spp.
The stinging hairs make this a difficult group to mistake; *handle only with gloves.* Erect, usually unbranched weeds. Note the minute greenish flowers in slender, forking clusters in the upper axils of the paired, toothed leaves. A number of species are found throughout our area; only 2 are shown; all are edible.
Use: Cooked green, soup, tea. The *young shoots* (while still only a few inches tall), and tender, pale green top *leaves* are excellent simmered for 10–15 min. in just enough water to cover and served with butter and lemon; the stinging qualities disappear upon cooking. Excellent added to soups or stews. To make a nourishing tea, boil the young shoots or leaves for several minutes, strain, and add lemon and sugar. Leaves contain vitamins A and C, as well as iron and protein. **Warning:** Do not handle with bare hands. If you do come into contact with the stinging hairs, a simple remedy is to scrub the affected area with crushed stems of one of the jewelweeds (pp. 78, 92).
SPRING (shoots)–SUMMER (tender leaves)
STINGING NETTLE, *U. dioica.* **Color pl. 9.** *Well-armed* with stinging hairs. Leaves ovate, with sharply toothed margins and heart-shaped bases. Stem 4-sided, hollow. 2–4 ft. (0.6–1.2 m). **Where found:** Roadsides, waste ground, light soils. Canada south to Ill. and Va. **Flowers:** June–Sept.
SLENDER NETTLE, *U. gracilis.* Resembles Stinging Nettle, but stinging hairs *sparse.* Leaves more slender; bases more rounded, less heart-shaped. 2–3 ft. (60–90 cm). **Where found:** Damp soil, thickets. Canada south to Minn., cen. N.Y., n. and w. New England; in mts. to W. Va. **Flowers:** July–Sept.

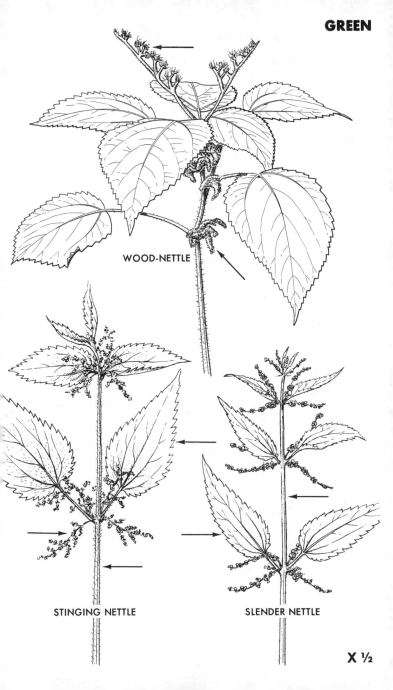

GREEN

WOOD-NETTLE

STINGING NETTLE

SLENDER NETTLE

X ½

CLUSTERS, TERMINAL AND IN UPPER AXILS

LAMB'S-QUARTERS, PIGWEED   **Tender leaves and**
Chenopodium album   **tips; seeds**
An erect, much-branched weed. Stems and undersides of leaves often mealy-white. Leaves somewhat fleshy, variable; upper leaves narrow, toothless; lower leaves roughly diamond-shaped, broadly toothed. Flowers small, unobtrusive, greenish, may turn reddish. Seeds tiny, black. 1–3 ft. (30–90 cm). Many similar species; most edible (see Mexican-tea, below). **Where found:** Waste ground. Throughout. **Flowers:** June–Oct.
Use: Cooked green, cereal, flour. The tender *leaves and tips* are excellent steamed or boiled for 10–15 min. Bulk greatly reduced after cooking. The highly nutritious *seeds* can be boiled to make a breakfast gruel, or ground into flour.
          SUMMER (leaves, tips); FALL–EARLY WINTER (seeds)

MEXICAN-TEA (leaf shown)   **Leaves (do not eat)**
Chenopodium ambrosioides
Resembles Lamb's-quarters (above), but note the *aromatic, wavy-toothed,* light green leaves, smelling of varnish. Flower clusters very leafy. 2–4 ft. (0.6–1.2 m). **Where found:** Waste ground. Northern U.S. south. **Flowers:** Aug.–Nov.
Use: Tea. The dried leaves are occasionally used to make tea.
Warning: Both this plant and JERUSALEM-OAK, *C. botrys* (sticky, oaklike leaves, smelling of turpentine), are typical of a small number of aromatic *Chenopodiums* that are covered with oil glands; do not eat their leaves.   SUMMER–FALL

ORACHE   **Tender leaves and tips**
Atriplex patula
One of several species of *Atriplex* that superficially resemble Lamb's-quarters (above). Leaves fleshy, arrow-shaped, with lower teeth pointing outward. Flowers in slender, interrupted clusters in leaf axils. 9–24 in. (22.5–60 cm). **Where found:** Saline or alkaline soil; saltmarshes, waste ground. Canada south to Kans., Mo., Ill., and N.C. **Flowers:** June–Nov.
Use: Cooked green. Steam or boil for 10–15 min.; serve like spinach. Reduces in bulk upon cooking.   SUMMER

STRAWBERRY-BLITE   **Tender leaves and**
Chenopodium capitatum   **tips, fruit**
Similar to Lamb's-quarters (above), but note the compact flower clusters which become bright red, fleshy, berrylike fruit. Leaves *triangular* to arrowhead-shaped, irregularly toothed. 6–24 in. (15–60 cm). **Where found:** Waste ground, recently burned clearings. Canada, n. U.S. (local). **Flowers:** June–Aug.
Use: Cooked green, fruit. Prepare the *tender parts* like Lamb's-quarters. The *fruit* are insipid but quite nutritious.
          SUMMER (leaves); LATE SUMMER (fruit)

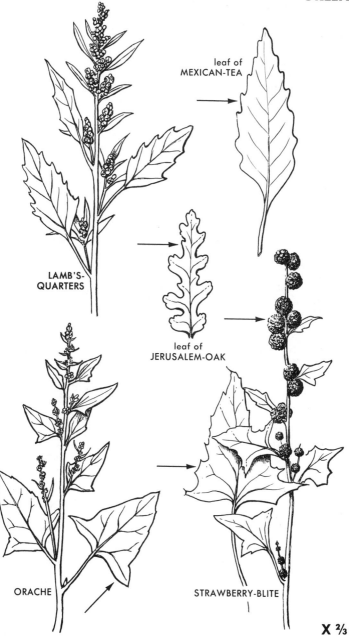

leaf of
MEXICAN-TEA

LAMB'S-
QUARTERS

leaf of
JERUSALEM-OAK

ORACHE

STRAWBERRY-BLITE

X ⅔

CLUSTERS, TERMINAL AND IN LEAF AXILS

AMARANTHS   **Tender leaves, seeds**
Amaranthus spp.   **Color pl. 8**

Coarse, hairy weeds with stout stems. Numerous species; 2 shown. GREEN AMARANTH, *A. retroflexus*. Leaves dull green, ovate to lance-shaped, long-stalked; flower clusters dense, bristly. 6–24 in. (15–60 cm). GREEN AMARANTH, *A. hybridus*. Taller; flower clusters slender, nodding. 2–6 ft. (0.6–1.8 m). **Where found:** Roadsides, fields, waste ground. Most of our area. **Flowers:** Aug.–Oct.
Use: Cooked green, salad, flour. The tender *leaves* can be boiled for 10–15 min., or added to salads. The tiny black *seeds* make a nutritious flour.

LATE SPRING–FALL (leaves); FALL–EARLY WINTER (seeds)

SORRELS   **Leaves and stems**
Rumex spp.

Note the slender clusters of small greenish or reddish flowers and *arrow-shaped leaves* that taste *sour*. 2 species shown.
Use: Salad, cold drink, cooked green. Outstanding raw, in salads or as a thirst-quenching nibble, or boiled 3–5 min.
Warning: See wood-sorrels, p. 104.   SPRING–LATE SUMMER
SHEEP or COMMON SORREL, *R. acetosella.* Small arrow-shaped leaves with spreading lobes and tiny flowers in sparse, branching spikes. 4–12 in. (10–30 cm). **Where found:** Thin fields, roadsides. Throughout. **Flowers:** June–Oct.
WILD SORREL, *R. hastatulus* (not shown). Almost identical to the preceding species but taller, with more conspicuous flower clusters. **Where found:** Sandy fields, roadsides. Coastal plain; N.C. south to Fla., Texas. **Flowers:** March–May.
GARDEN SORREL, *R. acetosa.* Coarser than Sheep Sorrel with much larger flowers and upper clusters. Upper leaves *embrace* stem. 6–24 in. (15–60 cm). **Where found:** Fields. Canada south to n. Penn., Vt. **Flowers:** June–Sept.

CURLED or YELLOW DOCK   **Young leaves**
Rumex crispus

Note the coarse leaves with *wavy margins* and dense heads of small greenish flowers or brownish seeds with heart-shaped wings. 1–4 ft. (0.3–1.2 m). (Numerous similar species of dock are in our area; all are more or less edible with varying overtones of bitterness and sourness). **Where found:** Fields, waste ground. Throughout. **Flowers:** March (south)–Sept. (north).
Use: Cooked green, salad. The very young leaves, before they become too bitter, can be added fresh to salads or boiled for 10–15 min. If the leaves are too bitter, 2 or 3 changes of water should be used. Rich in protein and vitamin A.   SPRING

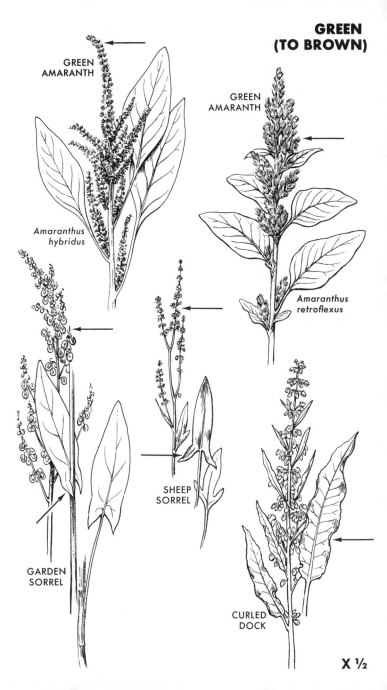

GREEN (TO BROWN)

GREEN AMARANTH

Amaranthus hybridus

GREEN AMARANTH

Amaranthus retroflexus

GARDEN SORREL

SHEEP SORREL

CURLED DOCK

X ½

CLUBLIKE SPADIX WITH MINUTE FLOWERS

JACK-IN-THE-PULPIT   **Dried corm**
Arisaema atrorubens
The flaplike spathe that curves over the club-shaped spadix is green or purplish brown and often striped. Leaves 1 or 2, on long succulent stalks, 3-parted. Fruit an egg-shaped cluster of scarlet berries. Corm walnut-sized or larger. 1–3 ft. (30–90 cm). **Where found:** Rich soil; woods, thickets, swamps. Se. Manitoba to Quebec, south to e. Kans., Tenn., and S.C. **Use:** Flour. The thinly sliced, *thoroughly dried* corms can be eaten as is, like potato chips, or ground into a pleasant cocoalike flour. **Warning:** Raw corms contain calcium oxalate. See Skunk Cabbage (below).   FALL–EARLY SPRING

SWEETFLAG, CALAMUS   **Rootstock, young shoots**
Acorus calamus   **Color pl. 6**
Yellow or yellow-green. See p. 62.   SPRING

ARROW ARUM, TUCKAHOE   **Dried fruit, dried**
Peltandra virginica   **rootstock**
A long — 4–7 in. (10–17.5 cm) — *pointed* spathe virtually conceals the slender spadix. Leaves large, *arrowhead-shaped*. Fruit a cluster of green or amber berries. Rootstock large, perpendicular, deeply buried. 12–18 in. (30–45 cm). A southern species, *P. sagittaefolia*, has a broader white spathe and crimson berries. **Where found:** Swamps, bogs. Great Lakes to s. Maine and south. **Flowers:** May–July.
Use: Flour, cooked vegetable (peas). **Warning:** Must be thoroughly dried before being eaten. See Skunk Cabbage (below) *and* Golden Club, p. 62.   LATE SUMMER–FALL (fruit);
FALL–EARLY SPRING (rootstock)

SKUNK CABBAGE   **Dried young leaves, dried**
Symplocarpus foetidus   **rootstock**
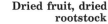
Crushed leaves with *skunklike odor*. Spathe shell-like, mottled with yellow-green or brown, enveloping a heavy round spadix. Leaves appearing after flowers; initially coiled into a small cone, later quite large. 1–3 ft. (30–90 cm). **Where found:** Wet woods, open swamps. S. Canada, n. U.S.; south locally in mts. to Tenn. and Ga. **Flowers:** Feb.–April.
Use: Cooked green, flour. The thoroughly dried *young leaves* are quite good reconstituted in soups or stews. The thoroughly dried *rootstocks* can be made into a pleasant cocoalike flour. **Warning:** Contains calcium oxalate crystals; eating the raw plant causes an intense burning sensation in the mouth. Boiling does not remove this property — only *thorough* drying. Also, do not confuse the young shoots with those of False Hellebore, p. 148.   EARLIEST SPRING (leaves);
FALL–EARLY SPRING (rootstock)

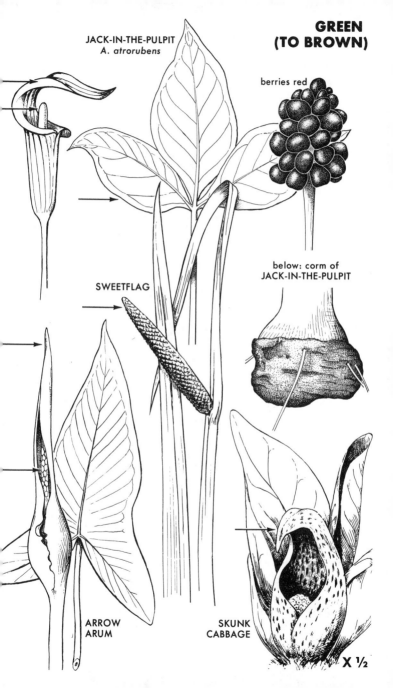

GREEN (TO BROWN)

JACK-IN-THE-PULPIT
A. atrorubens

berries red

below: corm of
JACK-IN-THE-PULPIT

SWEETFLAG

ARROW ARUM

SKUNK CABBAGE

X ½

SAUSAGELIKE HEAD OF
MINUTE FLOWERS

CATTAILS   **Young shoots and stalks, immature flower**
Typha spp.   **spikes, pollen, sprouts, rootstock**
Familiar plants often forming extensive stands in marshes.
Leaves erect, swordlike. Stems unbranched, stiff; topped by
compact, cylindrical heads of minute flowers; male flowers
above, golden when full of pollen, disappearing later; female
flowers in sausagelike heads below, green at first, then brown. 2
species shown. **Where found:** Fresh or brackish marshes, shallow water. Throughout. **Flowers:** May–July.
Use: Salad, asparagus, cooked vegetable, flour, pickle, potato.
This is possibly one of the best and certainly the most versatile
of our native edible plants. In early spring the young *shoots,*
which are easily pulled from the rootstocks, can be peeled to a
tender white core and eaten raw or cooked like asparagus
(boiled 15 min.). The young spring *stalks,* up to 2–3 ft. (60–
90 cm), can be gathered and prepared in much the same way.
In late spring the green *immature flower spikes* can be gathered
just before they erupt from their papery sheaths of leaves and
boiled for a few minutes. Served with butter and eaten like
corn on the cob, they make a unique wild vegetable. In early
summer, the flower spikes produce a heavy coat of bright yellow *pollen.* Large amounts of this pollen can be gathered by
shaking the heads in an open bag. After it is sifted through a
strainer, the pollen makes an excellent protein-rich flour when
mixed half-and-half with wheat flour. Be sure to dry the pollen
thoroughly before storing for future use. In late summer, small
horn-shaped *sprouts* form at the tip of the long rootstocks, and
remain through winter. These can be added to salads, or boiled
for 10 min. and served with butter. In earliest spring as they
begin to extend, but before they break through the surface of
the mud, these sprouts can be peeled, boiled briefly, and pickled
in hot vinegar. In addition, the starchy core at the base of each
sprout can be prepared like a potato. In late fall, winter, or
earliest spring, the shallowly buried *rootstocks* become well
filled with starch. To produce a good quality white flour, wash
the rootstocks thoroughly, peel off the outer covering to reveal
the starchy core, and crush the core in a pail of cold water,
separating the starch from the fibers. Remove the fibers, allow
the starch to settle, and pour off the water. When this washing
process has been repeated once or twice more, you will have
pure white flour that can be used immediately, or dried for
future use.   ALL YEAR (see text)
COMMON CATTAIL, *T. latifolia.* **Color pl. 3.** Broad-
leaved; male flowerhead *joins* female head. 3–9 ft. (0.9–2.7 m).
NARROW-LEAVED CATTAIL, *T. angustifolia.* Leaves narrower; *gap* between male and female flowerheads. 2–5 ft. (0.6–
1.5 m).

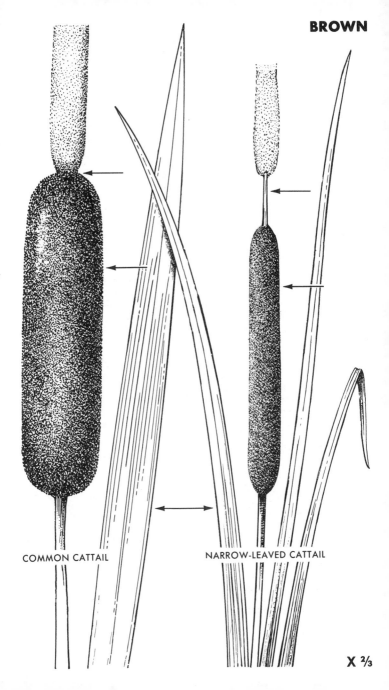

BROWN

COMMON CATTAIL

NARROW-LEAVED CATTAIL

X ⅔

MISCELLANEOUS BROWN FLOWERS

WILD GINGER Rootstock
Asarum canadense
A solitary, bell-shaped, red-brown flower with 3 spreading lobes sits on the ground between 2 stout, woolly leafstalks. Leaves in pairs; large, heart-shaped. Rootstock smells and tastes of ginger. Colonial. 6–12 in. (15–30 cm). **Where found:** Rich woods; near rocks. Minn. to Quebec and New Brunswick, south to Kans., Ark., Tenn., n. Ga. **Flowers:** April–May.
Use: Candy, spice. The long horizontal rootstocks lie just below ground and are easy to gather. To make a pleasant candy, cut the rootstocks into short sections, boil until tender (at least 1 hr.), simmer for another 20–30 min. in a rich sugar syrup, then separate and dry. The dried and crushed rootstocks can be substituted in recipes calling for commercial ginger.

EARLY SPRING–FALL

GROUNDNUT Tubers
Apios americana
A small, twining vine with compact, fragrant clusters of *maroon or lilac-brown,* pealike flowers in the leaf axils. Leaves light green, smooth; with 5–7 ovate, sharp-pointed leaflets. Root a long string of walnut-sized tubers. **Where found:** Thickets, streambanks, open woods. Minn., through Great Lakes area to New Brunswick, south to Texas and Fla. **Flowers:** June–Sept.
Use: Potato. Once they have been thoroughly washed, the small tubers are excellent boiled (about 20 min.), roasted, or fried in bacon fat; the taste is sweet and mildly turniplike. Eat the tubers while still hot; they become much less palatable when cold. ALL YEAR

FEVERWORT, TINKER'S-WEED Fruit
Triosteum perfoliatum

Note the paired leaves and the leafstalks with flaring margins that meet and surround the stem. Flowers in leaf axils, green or yellowish to dull purple-brown, bell-shaped, embraced by 5 long sepals. 2–8 hairy, orange to reddish-orange berries are clustered at each node in the stem; each berry contains 3 large nutlets and is crowned by 5 narrow sepals. 2–4 ft. (0.6–1.2 m). WILD COFFEE, *T. aurantiacum* (not shown) is similar, but its leaves taper to narrow bases. **Where found:** Woods, thickets. Minn., Wisc., Mich., and Mass., south to e. Kans. and Ga. **Flowers:** May–July. **Fruit:** Aug.-Oct.
Use: Coffee. Dry, roast, and grind the ripe berries. Add 1–2 tsp. of the ground berries for every cup of *cold* water and heat. When the mixture comes to a boil, remove it from the heat and allow it to steep for a few minutes; then drink. FALL

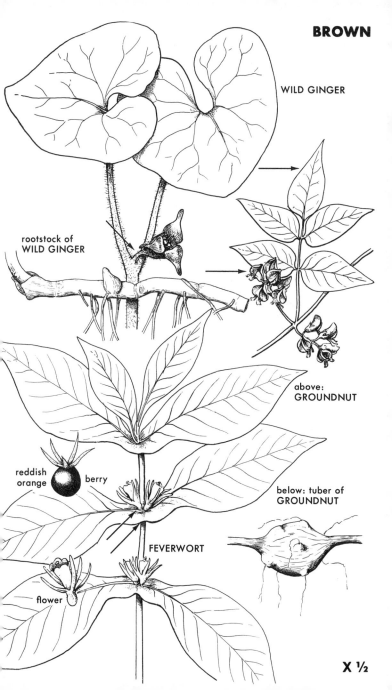

BROWN

WILD GINGER

rootstock of WILD GINGER

above: GROUNDNUT

reddish orange   berry

below: tuber of GROUNDNUT

FEVERWORT

flower

X ½

Color Plates

The 78 photographs in this section are organized according to habitat and depict characteristic edible species found in each locale (see pp. 241–285 for a description of each habitat and an explanation of their ordering). A fuller treatment of each species appears in the main text and is indicated by the page number(s) at the upper right-hand corner of every entry on the legend page.

Plate 1

SEASHORES

The shores along the Atlantic and Gulf coasts are the source of some truly exceptional wild foods.
See p. 242 for fuller details and available species.

SILVERWEED *Potentilla anserina* p. 70
The name Silverweed refers to the silvery-white undersides of the feather-compound leaves. The larger, more mature plants send down slender roots that make an excellent cooked vegetable, tasting vaguely like parsnips. FALL–EARLY SPRING

BEACH PEA *Lathyrus japonicus* pp. 122, 142
Clusters of small seedpods replace the showy purple or violet blossoms well before midsummer. The tiny peas contained in each pod resemble miniature garden peas and are excellent prepared in much the same way. SUMMER

WRINKLED ROSE *Rosa rugosa* pp. 106, 184
Roses are among the showiest and most widely recognized wildflowers. This is our largest species. In late summer it produces fat plum-sized fruit that are many times richer in vitamin C than oranges. SUMMER (petals);
LATE SUMMER–EARLY FALL (fruit)

MARSH MALLOW *Althaea officinalis* p. 108
Inadvertently brought to America by the early settlers, this velvety-leaved migrant from Europe can now be found along the margins of saltmarshes from Connecticut to Virginia. The long slimy roots were once a prime ingredient in the making of marshmallows. EARLY SUMMER (leaves);
SUMMER (flowerbuds); FALL–EARLY SPRING (roots)

SILVERWEED

BEACH PEA

WRINKLED ROSE

MARSH MALLOW

Plate 2

RUNNING WATER

The banks of springs and small, fast-flowing streams supply an abundance of edible greens during the early spring months, often before the first leaves appear on the trees.
See p. 245.

COMMON BLUE VIOLET *Viola papilionacea*　　p. 132
Even the most familiar things can have hidden virtues; so it is with violets. In addition to their usual role as a harbinger of spring, their leaves and flowers are potent sources of vitamins A and C.　　　　　　　　　　　　　Early Spring–Spring

SPOTTED TOUCH-ME-NOT *Impatiens capensis*　　p. 92
Although it makes a fine cooked green, Spotted Touch-me-not is best known as a medicine. Its stems and leaves produce a watery juice that is an extraordinarily effective remedy for Poison Ivy.　　　　　　　　　　Early Spring (shoots);
Summer (leaves, stems)

VIRGINIA MEADOW-BEAUTY *Rhexia virginica*　　p. 98
Meadow-beauty is too localized a plant to be assigned any general importance as a wild food, but when found in abundance it makes an excellent salad. Look for it growing in wet sandy soil in meadows or along sunny streams.　　Summer

WILD MINT *Mentha arvensis*　　　　　　pp. 54, 138
Because of their abundance and the ease with which they can be identified, the aromatic mints are the most widely used wild teas. Shown is a northern species commonly found in damp open soil on lake shores or along streams.　　　　Summer

COMMON ELDERBERRY　　　　　　　pp. 18, 172
Sambucus canadensis
By late summer, Common Elderberry's showy flower clusters are completely replaced by great nodding masses of tiny purple-black berries. Although somewhat rank and unpleasant-tasting raw, the berries make outstanding pastries and preserves when properly prepared.　　　　Summer (flowers);
Late Summer (fruit)

MARSH-MARIGOLD *Caltha palustris*　　　p. 70
Masses of dark green leaves and intensely colored blossoms make this one of the most spectacular spring wildflowers. The young leaves have been a traditional New England spring green since the days of the pilgrims.　　Early Spring–Spring

COMMON BLUE VIOLET

SPOTTED TOUCH-ME-NOT

VIRGINIA MEADOW-BEAUTY

WILD MINT

COMMON ELDERBERRY

MARSH-MARIGOLD

Plate 3

STILL WATER (Freshwater Ponds, Marshes)

If you do not mind getting your feet wet, areas of still or slow-moving fresh water make outstanding collecting sites. Edible species are abundant and relatively easy to gather throughout the year.

See p. 247.

COMMON CATTAIL *Typha latifolia*   pp. 158, 230
Probably no other plant, wild or domestic, is quite as versatile as Common Cattail. In addition to yielding food year-round, it provides the material for making torches, mattresses, rush seats, and flower arrangements.   ALL YEAR

WILD RICE *Zizania aquatica*   p. 228
Large stands of Wild Rice can be found throughout the Mississippi Valley and Great Lakes region. Although attempts have been made to grow it commercially, the best place to collect it is still in the wild.   MID–LATE SUMMER

BULLHEAD-LILY *Nuphar variegatum*   p. 60
The globelike yellow flower and platterlike floating leaves rise on separate stalks from a large, spongy, starch-filled rootstock. When properly prepared it has a sweetish taste and makes an excellent potato substitute.   LATE SUMMER–FALL (seeds);
FALL–EARLY SPRING (rootstock)

PICKERELWEED *Pontederia cordata*   p. 136
By late summer the showy spires of flowers have become tight clusters of irregular green fruit. These can be eaten out of hand, added to granola, or dried and ground into flour.
EARLY SUMMER (leaves); LATE SUMMER–FALL (fruit)

GREAT BULRUSH *Scirpus validus*   p. 230
Bulrush's tall naked stem and sparse tassle of chaffy brown spikelets form a startling contrast to most plants found in freshwater marshes. In areas where it is abundant, it can be put to almost as many uses as cattails (above). SPRING (shoots);
SUMMER (pollen); FALL (seeds);
FALL–EARLY SPRING (rootstock)

COMMON CATTAIL

BULLHEAD-LILY

PICKERELWEED

WILD RICE

GREAT BULRUSH

Plate 4

NORTHERN BOGS–ALPINE TUNDRA

Bogs and alpine tundra support many of the same edible species. Some of these are found in great abundance and are
highly prized as foods.
See Northern Bogs (p. 249) and Alpine Tundra (p. 283).

HIGHBUSH BLUEBERRY *Vaccinium corymbosum* p. 220
In addition to all their other uses, blueberries make an excellent
survival food. The ripe berries are not only abundant, but
nutritious and surprisingly filling as well.

MID–LATE SUMMER

CLOUDBERRY *Rubus chamaemorus* p. 32
Many consider this the most delicious of all far-northern berries. An alternate name, Bake-apple, refers to the applelike
flavor of the ripe fruit. (*Photograph by Dr. Steven Young:
Director, Center for Northern Studies; Wolcott, Vt.*)

SUMMER

NORTHERN FLY-HONEYSUCKLE pp. 78, 174
Lonicera villosa
This low honeysuckle produces small blue berries that suggest
blueberries in both taste and appearance. In Maine, it is called
Waterberry because of its habit of growing in the wet peaty
soils of swamps and bogs. (*Photograph by Cecil B. Hoisington:
Staff photographer, Vermont Institute of Natural Science.*)

SUMMER

LABRADOR-TEA *Ledum groenlandicum* pp. 18, 208
The small leathery leaves with their rolled edges and white or
rusty woolly undersides remain on the bush year-round. At one
time they were a popular substitute for oriental tea in the
far-northern states. ALL YEAR

LARGE CRANBERRY pp. 102, 222
Vaccinium macrocarpon
Cranberries are a typical component of the open bog mat, often
forming dense carpets over the sphagnum moss. If you know of
a nearby bog, you should be able to gather as much as you need.

FALL–WINTER

ALPINE SMARTWEED *Polygonum viviparum* p. 116
The thick tuberlike rootstock of this small arctic perennial
makes a delicious almond-flavored nibble. Unfortunately, it is
only found in the extreme northern limit of our range and even
then not very abundantly. (*Photograph by Dr. Steven Young:
Director, Center for Northern Studies; Wolcott, Vt.*)

SUMMER–FALL

HIGHBUSH BLUEBERRY

CLOUDBERRY

NORTHERN FLY-HONEYSUCKLE

LABRADOR-TEA

LARGE CRANBERRY

ALPINE SMARTWEED

Plate 5

SWAMPS

Swampy areas do not make ideal collecting sites, but they do
yield a variety of edible species found nowhere else.
See p. 251.

NORTHERN WILD-RAISIN *Viburnum cassinoides*   p. 178
Although this is a common shrub in the northern U.S., its fruit
is curiously neglected. The thin pulp surrounding the solitary
seeds is surprisingly good, tasting something like a cross between raisins and dates.   LATE SUMMER–FALL

RED CHOKEBERRY *Pyrus arbutifolia*   p. 220
Chokeberry's fruit suggest blueberries in size and shape, but
inside they are constructed more like an apple. Although little
used, they make excellent pies and preserves.   FALL

WILD CALLA *Calla palustris*   p. 22
The long creeping rootstock of this plant yields an edible starch
that at one time was used to make "missen" (famine) bread in
Scandinavia. Like Skunk Cabbage (p. 156), the raw plant has
an acrid peppery quality that is only removed by thorough and
prolonged drying.   FALL (seeds);
                        FALL–EARLY SPRING (rootstock)

COMMON SPICEBUSH *Lindera benzoin*   p. 208
An alternate name for this shrub is Wild Allspice. During the
Revolutionary War, its fruit was employed as a substitute for
allspice, and its twigs and leaves brewed into tea.

                                              ALL YEAR

SWEETGALE *Myrica gale*   p. 206
The narrow wedge-shaped leaves are highly aromatic, with a
fragrance suggestive of sage. Dried, they make an excellent tea.
                                              SUMMER

NORTHERN WILD-RAISIN

RED CHOKEBERRY

WILD CALLA

COMMON SPICEBUSH

SWEETGALE

Plate 6

WET MEADOWS

Low-lying grassy areas with wet or moist soil yield a variety of plants not normally found in higher, better-drained fields and pastures. Many of these wet meadow species are important wild foods.
See p. 253.

SWEETFLAG *Acorus calamus* pp. 62, 156
Sweetflag usually forms large colonies and can be readily identified by its yellow-green leaves and spicy-aromatic smell. Generations ago, it was a popular herb used as candy, salad, medicine, and even perfume. SPRING

SHRUBBY CINQUEFOIL *Potentilla fruticosa* pp. 70, 180
Except when in bloom, this plant is likely to be overlooked as just another low unobtrusive shrub. Its small 5- to 7-part leaves can be dried and made into a mild but pleasant tea. SUMMER

PEPPERMINT *Mentha piperita* pp. 118, 138
Little needs to be said about this plant. The familiar fragrance and hot taste of the leaves preclude misidentification; its uses are well known. SUMMER

WATER or PURPLE AVENS *Geum rivale* p. 94
The rootstock of this plant contains a substance not unlike chocolate in flavor. When boiled, with sugar and milk added, it makes an acceptable substitute for hot chocolate.
ALL YEAR

CANADA or WILD YELLOW LILY p.74
Lilium canadense
The true lilies, *Lilium* spp., are among our most spectacular native wildflowers. Although all produce large edible bulbs, most species are relatively scarce and their use as food plants is strongly discouraged. SUMMER

ANGELICA *Angelica atropurpurea* p. 40
The young stems are peeled, and then cooked like celery or made into an excellent candy. Great care should be exercised when collecting them, however; the foliage is somewhat similar to that of Water-hemlock (p. 42).
LATE SPRING–EARLY SUMMER

SWEETFLAG

SHRUBBY CINQUEFOIL

PEPPERMINT

PURPLE AVENS

CANADA LILY

ANGELICA

Plate 7

WASTE GROUND (1)

Waste places such as fallow fields, roadsides, and vacant lots yield an extraordinary variety of edible plants. Pictured here are four species commonly encountered along roadsides in summer.

See p. 256.

COMMON EVENING–PRIMROSE   p. 66
Oenothera biennis
The 4-petaled yellow blossoms open after dusk and often close again by midmorning. The first-year roots are delicious boiled like parsnips. They were one of our first edible wild plants to be exported back to Europe.

FALL (roots)–EARLY SPRING (roots, leaves)

DAY-LILY *Hemerocallis fulva*   p. 92
While Day-lily is often planted around homes for its large showy flowers which bloom and wither in a day, most people are surprised to discover that it also furnishes a year-round supply of food. The shoots, flowerbuds, flowers, and tubers are all edible and delicious.   ALL YEAR (tubers);
EARLY SPRING (shoots); SUMMER (buds, flowers)

CHICORY *Cichorium intybus*   pp. 58, 144
Chicory is another of those plants that have followed man's migration to the New World. Like Common Dandelion (p. 84) it is used extensively as a spring green and coffee substitute.

EARLY SPRING (leaves); FALL–EARLY SPRING (roots)

YARROW *Achillea millefolium*   p. 38
Yarrow's finely-cut, fernlike leaves are usually employed as an herbal medicine, reportedly curing such diverse ailments as cuts, colds, baldness, and an unhappy love life. Some people, however, just like them brewed into tea.   SUMMER

COMMON EVENING-PRIMROSE

DAY-LILY

CHICORY

YARROW

Plate 8

WASTE GROUND (2)

Barnyards and lawns are particularly productive sites for the forager. The species listed below are some of the ones typically found in such places.

See p. 256.

POKEWEED *Phytolacca americana* p. 46
The young leafy shoots have long been a popular spring green, but great care must be taken in their collection and preparation. The roots, the mature stems and leaves, and the seeds are all dangerously poisonous. SPRING

WINTER CRESS *Barbarea vulgaris* p. 64
Winter Cress is one of the few greens available during the colder months of the year, sending up tender new leaves during winter warm spells and in earliest spring. Although initially mild tasting, the leaves rapidly become bitter and are no longer usable by the time the flowers appear.
LATE WINTER–EARLY SPRING (leaves); SPRING (flowerbuds)

GILL-OVER-THE-GROUND *Glechoma hederacea* p. 140
This low-growing relative of Catnip (pp. 54, 140) was at one time employed as a bitter tonic (using the green leaves). Like those of Yarrow (p. 38), the dried leaves are sometimes made into tea. SPRING–SUMMER

GREEN AMARANTH *Amaranthus retroflexus* p. 154
In Central America amaranths are cultivated and their heavy crop of tiny seeds ground into flour. Our own species are not so heavily laden with seeds, but they can be used similarly.
LATE SPRING–FALL (leaves); FALL–EARLY WINTER (seeds)

COMMON DANDELION *Taraxacum officinale* p. 84
Dandelion belongs to a category of plants known as cosmopolitan weeds—species that thrive in disturbed soil and have dogged man's footsteps around the globe. A surprisingly high percentage of these are edible, and some have even been cultivated as major food crops. EARLY SPRING (leaves, buds);
SPRING–EARLY SUMMER (flowers);
FALL–EARLY SPRING (roots)

CHEESES *Malva neglecta* pp. 32, 108, 134
Cheeses is a low, diminutive mallow usually found creeping around barnyards. Its name comes from the flattened green fruit which look like miniature wheels of cheese.
SPRING (leaves); SUMMER (leaves, fruit)

POKEWEED

WINTER CRESS

GILL-OVER-THE-GROUND

GREEN AMARANTH

COMMON DANDELION

CHEESES

Plate 9

WASTE GROUND (3)

The four roadside species below make their appearance in summer. Each of them yields a number of outstanding dishes. See p. 256.

STINGING NETTLE *Urtica dioica* p. 150
Despite their plain appearance and stinging (urticating) hairs, nettles are outstanding wild foods. When properly prepared, the early leaves are not only delicious but extraordinarily rich in vitamins A and C, as well as protein, chlorophyll, phosphorus, and iron. SPRING (shoots); SUMMER (tender leaves)

COMMON BURDOCK *Arctium minus* p. 126
Burdock's rank leaves and tenacious burs are a familiar part of most disturbed areas. Despite its reputation as an obnoxious weed, it is the source of several palatable foods.
SPRING (leaves); EARLY SUMMER (roots)

COMMON SUNFLOWER *Helianthus annuus* p. 88
Common Sunflower is a frequent roadside weed in the midwestern prairie states. The seeds are slightly smaller than those sold commercially, but are just as good.
LATE SUMMER–FALL

FIREWEED *Epilobium angustifolium* p. 98
Fireweed gets its name not from its spire of rose-purple blossoms but from its tendency to grow in places that have recently been burned. Early in the year, the young shoots can be boiled and eaten like asparagus; later, the mature leaves can be dried and used to make tea. LATE SPRING (shoots);
SUMMER (leaves)

STINGING NETTLE

COMMON BURDOCK

COMMON SUNFLOWER

FIREWEED

Plate 10

OPEN FIELDS

Tall herbaceous plants may give an open field the appearance
of a wild flower garden. This is a rapidly changing, unstable
habitat whose dominant feature is its grasses. The advent of
the taller herbs and shrubs, however, begins a shading-out
process that causes the grass to disappear and the field to
eventually pass into forest. Open fields provide a wide variety
of salads and cooked greens during the spring and summer.
See p. 262.

SPIDERWORT *Tradescantia virginiana* p. 130
Intensely colored flowers and long graceful leaves make
Spiderwort one of the more eye-catching wildflowers. The
young leaves can be added to salads and the flowers made into
candy. SPRING (leaves);
LATE SPRING–EARLY SUMMER (flowers)

COMMON MULLEIN *Verbascum thapsus* p. 72
The velvety gray-green leaves and clublike flowerheads of
Common Mullein are a familiar feature of pastures and road-
sides in the East. A delicate tea brewed from the dried leaves
was at one time a popular herbal remedy for lung ailments.
SUMMER

COMMON STRAWBERRY *Fragaria virginiana* p. 30
The small juicy berries produced by the wild strawberries are
tedious to gather, but well worth the effort; their flavor is far
superior to that of domestic strawberries. Many consider them
to be the best of the wild fruits. SUMMER

BULL THISTLE *Cirsium vulgare* p. 126
With their stiff bristly spines, thistles seem unlikely food
plants. However, when properly prepared, their basal leaves,
young stems, and first-year roots all provide palatable dishes.
SPRING–FALL

COMMON MILKWEED *Asclepias syriaca* p. 112
Despite its bitter milky juice, this is a versatile and outstanding
food plant. The young shoots, flowerbuds, flowers, and young
seedpods all make delicious dishes. SPRING (shoots, leaves);
SUMMER (buds, flowers, pods)

SPIDERWORT

COMMON MULLEIN

COMMON STRAWBERRY

BULL THISTLE

COMMON MILKWEED

Plate 11

OLD FIELDS

Old fields represent a midpoint in the transition between an open field and a thicket. The biennial and perennial herbs that characterize an open field have, for the most part, disappeared. In their place appear a variety of edible shrubs and small trees.
See p. 265.

BLACK LOCUST *Robinia pseudo-acacia*   p. 184
Black Locust's pendulous flower clusters appear in late spring and fill the air with their fragrance. They make excellent fritters, tasting much the way they smell.   SPRING

COMMON BARBERRY *Berberis vulgaris*   p. 192
Although this plant has been eliminated from much of the West because of its role as an intermediate host for wheat rust, it is still a familiar sight in old fields and roadside thickets throughout the Northeast. The scarlet berries are extremely tart and make an outstanding wild jelly.   FALL

SMOOTH SUMAC *Rhus glabra*   p. 186
The tiny red fruit of the sumacs are covered with acidic hairs and yield a drink remarkably similar to pink lemonade when soaked in cold water. They must be gathered during periods of good weather, however, as heavy or prolonged rains will wash out most of the acid.   SUMMER

COMMON BLACKBERRY   pp. 30, 184
Rubus allegheniensis
There are over 200 species of blackberries, raspberries, and dewberries in the East; this is our commonest blackberry. Not only does it produce delicious fruit, but its young shoots can be added to salads and its leaves used to make tea.
SPRING (shoots); SUMMER (leaves, fruit)

BLACK LOCUST

COMMON BARBERRY

SMOOTH SUMAC

COMMON BLACKBERRY

Plate 12

THICKETS

Although thickets can supply food year-round, they are most productive in the late summer or early fall when the fruit-bearing shrubs and trees come into season.

See p. 268.

CARRION-FLOWER *Smilax herbacea*   p. 148
Despite its unattractive name (the flowers smell of carrion), this is an excellent food plant. The vigorous spring shoots are prepared like asparagus and have a delicate flavor rivaled by few other vegetables.   EARLY SPRING (shoots);
ALL YEAR (rootstock)

HIGHBUSH-CRANBERRY *Viburnum trilobum*   p. 178
Although Highbush-cranberry is not related to cranberries, its fruit have many of the same uses. They ripen to a bright red in the late summer or early fall and are often held on the bush until early spring.   LATE SUMMER–FALL

PURPLE-FLOWERING RASPBERRY   pp. 106, 212
Rubus odoratus
The shallow, cup-shaped fruit that follow the blossoms are tart and somewhat dry. They can be eaten raw or made into a delicious jelly.   SUMMER

COW-PARSNIP *Heracleum maximum*   p. 40
Huge leaves and great umbrellalike flower clusters make this one of the easiest members of the Carrot Family to identify. The raw plant has a disagreeable taste, but becomes quite palatable when properly prepared.
SPRING (stems, leafstalks);
SUMMER (seeds); FALL–EARLY SPRING (roots)

JERUSALEM ARTICHOKE *Helianthus tuberosus*   p. 88
This outstanding food plant was originally native to the Midwest, but has now escaped from cultivation and become thoroughly naturalized in the East. In the fall it produces large crisp tubers that can be eaten raw or prepared like potatoes.
FALL–EARLY SPRING

SIBERIAN CRAB *Pyrus baccata*   p. 216
Crabapples are usually too hard and sour to be eaten fresh, but they make delicious jellies and preserves. Pictured is an Asian species that occasionally escapes from cultivation and spreads to nearby thickets and clearings.   FALL

CARRION-FLOWER

HIGHBUSH-CRANBERRY

PURPLE-FLOWERING RASPBERRY

COW-PARSNIP

JERUSALEM ARTICHOKE

SIBERIAN CRAB

Plate 13

DRY OPEN WOODS

Sunny woods with sandy, well-drained soils are not quite the barren habitats they often appear. Some of our more interesting edible plants can be found in such places.
See p. 272.

SULPHUR SHELF *Polyporus sulphureus* p. 238
The bright yellow and orange hues of this bracket fungus make it impossible to misidentify. An alternate name, Chicken-of-the-woods, alludes to its flavor when cooked.
LATE SUMMER–FALL

SAW PALMETTO *Serenoa repens* p. 170
Gathering the spikelike terminal buds of this plant is a tedious and often painful process, but well worth the effort. The tender base of the bud, the palm-heart, is a truly exceptional wild food.
ALL YEAR

NEW JERSEY TEA *Ceanothus americanus* pp. 18, 192
The dried leaves yield a tea somewhat similar to oriental tea in flavor but without the caffeine. In fact, during the Revolutionary War it was a popular substitute for the real article.
SUMMER

SWEET GOLDENROD *Solidago odora* p. 90
The delicate aniselike odor of the crushed leaves distinguishes this from any other goldenrod. The fresh or dried leaves and flowers make a delightful tea. SUMMER

WHITE OAK *Quercus alba* p. 204
Although generally ignored as a food source today, acorns were a "staff of life" for many primitive peoples. Properly prepared, they not only rival chestnuts in flavor but contain 37 percent fat and 8 percent protein as well. EARLY FALL

SULPHUR SHELF

SAW PALMETTO

NEW JERSEY TEA

SWEET GOLDENROD

WHITE OAK

Plate 14

MOIST WOODS

The rich, moist soils associated with some deciduous forests are extraordinarily productive. A host of edible herbs can be gathered on the forest floor in early spring and nuts harvested from the trees in fall.

See p. 275.

TROUT-LILY *Erythronium americanum* p. 74
The mottled leaves and bell-like flowers are a familiar sight in the moist alluvial soils bordering streams and rivers. Both the leaves and bulbs are edible but should be sampled sparingly at first, as they may be mildly emetic. EARLY SPRING

RED or PURPLE TRILLIUM *Trillium erectum* p. 96
Trilliums' newly emerging shoots can be used as either a salad or cooked green. However, the plants' relative scarcity and great beauty make it inadvisable to gather them except in emergency. EARLY SPRING

CAROLINA SPRING-BEAUTY pp. 32, 104
Claytonia caroliniana
The small, irregularly shaped corms of this delicate spring flower are delicious cooked, tasting somewhat like chestnuts. Unfortunately, digging them up is often a tedious business and should be attempted only in areas where the plant is very abundant. SPRING

WILD OATS *Uvularia sessilifolia* p. 76
Once the leafy heads have been removed, Wild Oats' slender shoots can be prepared like asparagus. The other species of *Uvularia,* bellworts, are also edible, but are not usually abundant enough to warrant collection. EARLY SPRING

COMMON MOREL *Morchella esculenta* p. 238
Although generally ignored as a possible food source, several common and readily identified wild mushrooms are available to the enterprising forager. This is probably the most prized species, usually making its appearance at about the time the leaves first appear on the oaks. SPRING (usually May)

TOOTHWORT *Dentaria diphylla* pp. 28, 100
The small, four-petaled, white or pink flowers make their appearance in early spring and often carpet large areas. The slender stems rise from a peppery-pungent rootstock that can be grated and prepared like horseradish. SPRING

TROUT-LILY

RED or PURPLE TRILLIUM

CAROLINA SPRING-BEAUTY

WILD OATS

COMMON MOREL

TOOTHWORT

Plate 15

SPRUCE–FIR (BOREAL) WOODS

Spruce-fir woods enter our range only at its northern limit. The forest floor yields a limited number of edible species, but these are usually quite abundant.

See p. 281.

BUNCHBERRY *Cornus canadensis*   p. 20
Bunchberry's vivid fruit clusters are easy to spot against the gentle greens and browns of the forest floor in late summer. Although essentially tasteless, the berries are plentiful and nutritious.   LATE SUMMER–EARLY FALL

CREEPING SNOWBERRY   pp. 36, 222
Gaultheria hispidula
Normally this is an unobtrusive little plant creeping over moss-covered rocks or logs. Only the wintergreen odor of the crushed leaves and an occasional heavy crop of small white berries hint at its potential as a food plant.
                    ALL YEAR (leaves); LATE SUMMER (fruit)

CLINTONIA *Clintonia borealis*   pp. 74, 148
Clintonia leaves must be eaten when they first appear. As they unfurl, their cucumber flavor quickly becomes harsh and unpleasant, and they are no longer edible by the time the flowers appear.   EARLY SPRING

COMMON WOOD-SORREL *Oxalis montana*   pp. 30, 104
The pleasantly sour taste of the leaves is due to small amounts of a poison, oxalic acid. Although the leaves of this plant are quite safe if eaten in moderate amounts, it is best to remember that they contain the same chemical that makes rhubarb leaves deadly.   SPRING–SUMMER

PARTRIDGEBERRY *Mitchella repens*   pp. 36, 102, 174
The twin flowers produce a single dry red berry that remains on the plant until the next spring. Although tasteless, the berries make an easily recognizable emergency food. (*Photograph by Cecil B. Hoisington: Staff photographer, Vermont Institute of Natural Science.*)   FALL–EARLY SPRING

NORTHERN MOUNTAIN-ASH *Pyrus decora*   p. 188
Despite their attractive appearance, the berries pictured are too bitter to eat. Only in the late fall or winter, when thoroughly ripe and mellowed by frost, does their flavor become pleasantly tart.   FALL–EARLY SPRING

BUNCHBERRY

CREEPING SNOWBERRY

CLINTONIA

COMMON WOOD-SORREL

PARTRIDGEBERRY

NORTHERN MOUNTAIN-ASH

Woody Plants
(Trees, Shrubs, Vines)

EVERGREENS WITH FLAT NEEDLES
AND SPRAYS

EASTERN HEMLOCK　　　　　**Young needles, inner bark**
Tsuga canadensis

Note the loose, irregular, feathery outline. Needles suggest Balsam Fir (below) with *whitened* undersides and flat sprays, but shorter — to $\frac{1}{2}$ in. (1.3 cm) — attached to twigs by *slender stalks.* Twigs *rough* after needles removed. Cones small — to 1 in. (2.5 cm) — brown, pendent. Bark rough, dark. Up to 70 ft. (21 m). **Where found:** Mostly hilly or rocky woods. S. Canada south to Ind., Ky., Md.,; in mts. to Ala., n. Ga.
Use: Tea, flour. The light green *young needles* make a pleasant tea rich in vitamin C. In emergencies the *inner bark* can be used like that of Balsam Fir (below).　　　　　SPRING

AMERICAN YEW　　　　　　　　　　**Poisonous**
Taxus canadensis
A low, straggling evergreen shrub with flat, pointed needles up to 1 in. (2.5 cm) long in flat sprays somewhat like those of Balsam Fir (below). *Needles green on both sides,* stalked. Twigs smooth after needles removed. Note the juicy, *red ber-rylike fruit* with the single hard seed. Height 3 ft. (90 cm).
Where found: Moist woods. Canada south in uplands to ne. Iowa, ne. Ky., w. Va., New England.
Warning: Although the sweet pulp around the seeds is edible, the seeds themselves and the wilted foliage contain the heart-depressing alkaloid, taxine.

BALSAM FIR　　　　　　　　　**Inner bark, pitch**
Abies balsamea
A steeple-shaped tree. Needles up to $1\frac{1}{4}$ in. (3.1 cm) long, with 2 *whitish lines beneath;* stalkless, with enlarged *saucerlike bases;* arranged in 2-sided flat sprays. Bare twigs smooth. Cones 1–3 in. (2.5–7.5 cm) long, purplish to green, *erect.* Bark of younger trees smoothish, with numerous resin-filled blisters. Up to 60 ft. (18 m). **Where found:** Cool moist woods. Canada south to ne. Iowa, Mich., n. Ohio, New England; in mts. to W. Va. and Va.
Use: Emergency food, flour. Although disagreeable tasting, the *pitch* contained in the blisters in the smooth bark is a concen-trated food and should be remembered in times of need. The soft *inner bark,* or cambium, can be dried and ground into meal and mixed with flour to extend supplies in times of emergency; the taste of the mixture is unattractive but nourishing.
　　　　　SPRING (inner bark); ALL YEAR (pitch)

EASTERN HEMLOCK

silhouette of EASTERN HEMLOCK

red

AMERICAN YEW

silhouette of BALSAM FIR

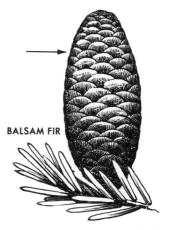

BALSAM FIR

X ⅔

BUNDLES OF 2 OR MORE NEEDLES

 TAMARACK, AMERICAN LARCH   **Young shoots,**
Larix laricina   **inner bark**
A medium to tall, pointed, cone-bearing tree of cold northern
swamps and bogs. Needles up to 1 in. (2.5 cm) long, soft, slender, pale green; in dense brushlike tufts on short *warty spurs*
along twigs; deciduous, turning yellow and then dropping in
fall. Cones ovoid, erect, up to ¾ in. (1.9 cm) long. Bark grayish
to brown, thin, flaking off in small scales. Height 40–80 ft.
(12–24 m); diameter 1–2 ft. (30–60 cm). **Where found:**
Swamps, bogs. N. Canada south to Minn., n. Ill., n. W. Va.,
n. Md., n. N.J.
Use: Cooked vegetable, flour. The tender new *shoots* (cooked)
and *inner bark* are nutritious and can be used in emergencies;
see Balsam Fir (p. 164).   SPRING

PINES   **Young shoots, inner bark,**
Pinus spp.   **young male cones, needles**
Widespread, familiar, cone-bearing, evergreen trees with clusters of long, slender needles along the twigs. Needle clusters
bound at base into bundles; each bundle with 2–5 needles.
Male cones (catkins) small, pollen-producing; female cones
larger, woody, seed-bearing. All pines edible; some better than
others. **Where found:** Dry to moist soil. One or more species
throughout. WHITE PINE, *P. strobus,* with 5 needles in each
bundle, shown.
Use: Candy, cooked vegetable, flour, tea. The tender new
shoots, stripped of their needles and peeled, can be made into
an acceptable candy using the method described for Wild Ginger (p. 96). In emergencies, the firm *young male cones* can be
boiled, and the *inner bark* can be made into flour (see Balsam
Fir, p. 164); piny flavored but highly nutritious. Chopped fine
and steeped in hot water, the fresh *needles* make an aromatic
tea rich in vitamins A and C; the light green needles from the
spring shoots make the best tea, but older needles can be used
as well.   SPRING (shoots, inner bark, male cones);
                                    ALL YEAR (needles)

deciduous

TAMARACK

numerous soft needles

silhouette of TAMARACK

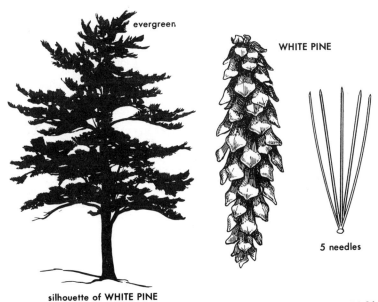

evergreen

WHITE PINE

5 needles

silhouette of WHITE PINE

X 2/3

MISCELLANEOUS EVERGREEN TREES AND SHRUBS

BLACK CROWBERRY   **Fruit**
Empetrum nigrum
A low, mat-forming shrub with tiny pink flowers tucked in the
axils of short — $\frac{1}{4}$ in. (6 mm) — somewhat elliptic, needlelike
leaves. Fruit a pea-sized, juicy, black berry. **Where found:**
Open peaty soil, tundra. Arctic south to northern edge of U.S.
Flowers: July–Aug. **Fruit:** July–Nov. (on through winter).
Use: Fresh or cooked fruit. The flavor of the berries improves
after freezing, or when cooked with a little lemon juice and
served with cream and sugar. Add the fresh berries to pancakes
and muffins.   LATE SUMMER–WINTER

GROUND JUNIPER   **Fruit**
Juniperus communis var. *depressa*
A flattened shrub. Needles $\frac{1}{2}$ in. (1.3 cm) long, sharp-pointed,
hollow, 3-sided, *whitened above; in whorls of 3.* Fruit a small,
hard, blue-black berry covered with whitish powder. 1–4 ft.
(0.3–1.2 m). **Where found:** Rocky, poor soil. Canada south to
Minn., Ill., Va., in mts. to Tenn., Ga.
Use: Seasoning. The crushed berries make a fine seasoning
with veal or roast of lamb.   SUMMER–FALL

SPRUCES   **Young shoots, inner bark, pitch**
Picea spp.
Steeple-shaped trees of cold climates with short, stiff, sharp-
pointed, 4-sided needles that grow all around the twigs. Twigs
rough when needles removed. Cones brown, woody, drooping.
Bark rough and dark. 1 species shown.
Use: Cooked vegetable, flour, chewing gum. In emergencies,
the tender leading *shoots* can be stripped of their needles and
boiled, and the *inner bark* can be used like that of Balsam Fir
(p. 164). The condensed and hardened sap, or *pitch,* is a familiar north woods substitute for chewing gum.
        SPRING (shoots, inner bark); ALL YEAR (pitch)
RED SPRUCE, *P. rubens.* Twigs hairy. Needles dark green or
yellow-green. $\frac{1}{2}$ in. (1.3 cm) long. Height 60–70 ft. (18–21 m);
diameter 1–3 ft. (30–90 cm). **Where found:** Well-drained soils.
S. Quebec to Nova Scotia, south to ne. Penn., e. N.Y., n. N.J.; in
mts. to e. Tenn., w. N.C.
BLACK SPRUCE, *P. mariana* (not shown). Somewhat
smaller with shorter needles and cones. **Where found:** Bogs,
wet soil. N. Canada, northern edge of U.S.

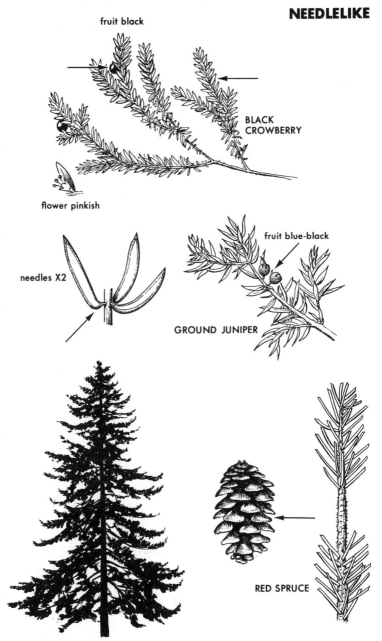

fruit black

BLACK CROWBERRY

flower pinkish

needles X2

fruit blue-black

GROUND JUNIPER

RED SPRUCE

silhouette of RED SPRUCE

X ⅔

EVERGREEN SHRUBS; DAGGERLIKE BLADES:
PALMS AND YUCCAS

SAW PALMETTO   **Terminal bud**
Serenoa repens   **Color pl. 13**
Our most abundant native palm, often forming dense stands.
Usually a low shrub with thick, branched, creeping or horizontal stems; occasionally erect, reaching tree-size. Leaves stiff,
erect, *fan-shaped;* 1–3 ft. (30–90 cm) wide; with numerous
narrow, bladelike segments. Leafstalks usually *longer* than
blades, armed with sharp *recurved spines.* Flowers ivory-white;
in large, fragrant, plumelike clusters. Fruit oblong, black when
ripe, ½–1 in. (1.3–2.5 cm) long. Height usually 3–7 ft. (0.9–
2.1 m). **Where found:** Sandy soils; pine forests, prairies,
dunes. E. La. to Fla., north to S.C. **Flowers:** May–July.
Fruit: Oct.–Nov.
Use: Salad, cooked vegetable. New leaves develop one at a
time from a spikelike terminal bud at the top of the stem. The
base of this bud, the palm-heart, is firm-textured, yet tender,
and makes an excellent salad or cooked vegetable. **Note:** Although removing the terminal bud kills the plant, palmettos
are tremendously abundant — virtual weeds in the deep South.
ALL YEAR

YUCCAS   **Petals; ripe fruit (Spanish Bayonet)**
Yucca spp.
Woody-stemmed evergreen plants bristling with stiff, swordlike
or daggerlike leaves. Flowers 6-petaled, waxy-white, showy, in
large terminal clusters. Fruit picklelike in shape, roughly 6-
sided. Numerous southwestern species; 2 from our area shown.
Use: Salad, cooked vegetable. The large *petals* make an interesting addition to salads. Pulp from the *ripe fruit* of Spanish
Bayonet (and other succulent-fruited species) can be eaten
cooked: halve the fruit, scrape out the seeds and fiber, wrap in
aluminum foil, and bake for 30 min. at 350°F (173°C).
SUMMER

SPANISH BAYONET, *Y. aloifolia.* Tall, erect. Stems thick,
clublike; often covered with downward-pointing old leaves, and
with a bristly pompon of newer leaves at the summit. Leaves
daggerlike, minutely saw-toothed on margins. Flower clusters
dense. Ripe fruit purplish, pulpy. Up to 15 ft. (4.5 m). **Where
found:** Sandy woods, dunes. Coastal plain; Ala. to Fla., north
to N.C. **Flowers:** May–June.
YUCCA, BEAR-GRASS, *Y. filamentosa.* Leaves close to
ground, in a *basal* rosette; somewhat flexible. Note loose
threads on leaf margins. Flowers in a large loose cluster atop a
slender stalk 3–5 ft. (0.9–1.5 m) high. Ripe fruit dry capsules.
Where found: Sandy woods, clearings, old fields. Coastal
plain from s. N.J. south; a garden escape elsewhere. **Flowers:**
May–Sept.

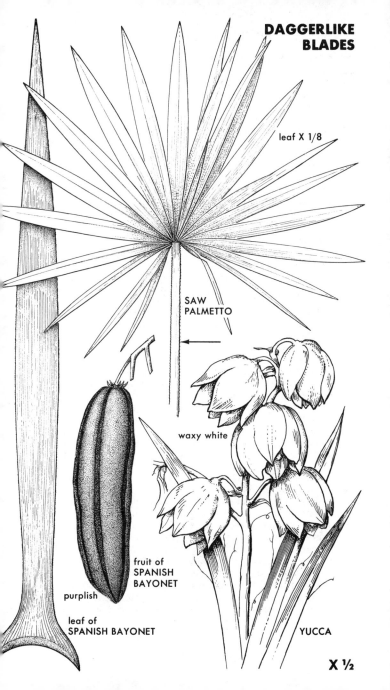

DAGGERLIKE BLADES

leaf X 1/8

SAW PALMETTO

waxy white

fruit of SPANISH BAYONET

purplish

leaf of SPANISH BAYONET

YUCCA

X ½

COMMON ELDERBERRY   Flower clusters,
Sambucus canadensis   Color pl. 2   ripe fruit (only)
A common erect shrub. Leaves divided into 5–11 coarsely-toothed, elliptic leaflets; 4–11 in. (10–27.5 cm) long. Twigs stout, soft, with a *thick white pith.* Flowers small, white, in dense *flat-topped clusters.* Berries tiny, juicy, *purple-black.* 3–13 ft. (0.9–3.9 m). **Where found:** Wet or damp rich soils; streambanks, thickets, roadside ditches. Canada south to Texas, Ga. **Flowers:** June–July. **Fruit:** Aug.–Oct.

Use: Fritters, jelly, cold drink, fruit. The large *flower clusters* can be dipped in batter and fried in oil to make fritters. Although the fresh *berries* are rank smelling and mildly unpleasant tasting, they are excellent when properly prepared. They make an outstanding jelly when mixed with the juice of one of the more tart fruits such as green apples, crabapples, or half-ripe grapes. Add one cup of water for every quart of berries, simmer for 10–15 min., mash, and simmer for another 10 min., then strain through several layers of cheesecloth. Prepare the juice of the tart fruit you choose in a similar manner. Combine the two juices, sweeten to taste and heat again until ready to jell. Elderberries do not contain their own pectin, so if you do not add at least as much of the tart juice as elderberry juice, you will need to use commercial pectin. Elderberry juice can also be chilled and used as a cold drink. Elderberries are easily dried by separating them from their stalks and placing them in the sun on flat trays; not only does the drying remove the rank odor and taste, but it allows one to store them for winter use. Once the dried berries have been reconstituted in boiling water, they are excellent added to muffins, mixed in fruit stews, or made into pie filling. Berries extremely rich in vitamin C; also contain vitamin A, calcium, iron, and potassium. **Warning:** Although flowers and ripe fruit are perfectly edible, the roots, stems, leaves, and unripe berries can cause nausea, vomiting, and diarrhea.   SUMMER (flowers); LATE SUMMER (fruit)

HORSECHESTNUT   Poisonous
Aesculus hippocastanum
A large tree. Leaves palm-shaped, with 7–9 wedge-shaped, toothed leaflets; 4–15 in. (10–37.5 cm) long. Flowers white, in showy upright spikes. Note the 3-part, thorny husks and large shiny-brown nuts. Up to 75 ft. (22.5 m). Several similar native species, buckeyes, *Aesculus* spp., with 5 leaflets and yellow or red flowers, are also poisonous; fruit of SWEET BUCKEYE, *A. octandra,* shown. **Where found:** Planted around homes. Much of our area. **Flowers:** May. **Fruit:** Sept.–Oct.
Warning: All parts contain a dangerous glycoside. Nuts should not be eaten, even after thorough soaking.

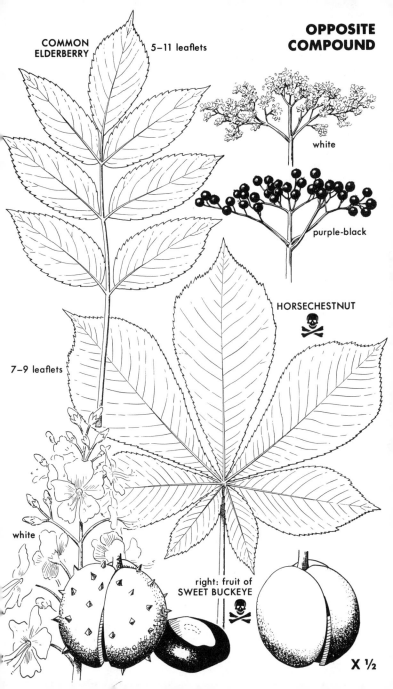

OPPOSITE COMPOUND

COMMON ELDERBERRY

5-11 leaflets

white

purple-black

7-9 leaflets

white

HORSECHESTNUT ☠

right: fruit of SWEET BUCKEYE ☠

X ½

MISCELLANEOUS SHRUBS
WITH TOOTHLESS LEAVES

PARTRIDGEBERRY   **Fruit**
Mitchella repens   **Color pl. 15**
Note the small, paired, roundish evergreen leaves along the slightly woody, creeping stem; leaves often variegated with whitish lines. Pink or white 4-petaled flowers, in *twinlike union* terminate the stem. Fruit a bright red berry. **Where found:** Moist woods. Most of our area. **Flowers:** June–July. **Fruit:** July–winter.
Use: Nibble, salad. The small red berries are dry, seedy, and tasteless, but are quite edible raw and may be used as a colorful addition to salads. The berries often remain on the stems over winter.   Fall–Early Spring

NORTHERN FLY-HONEYSUCKLE   **Fruit**
Lonicera villosa   **Color pl. 4**
A small, erect honeysuckle with *densely hairy* oval leaves and downy twigs. Twigs with a solid pith. Leaves 1–2½ in. (2.5–6.3 cm) long. Yellowish 5-petaled flowers are on short stalks. Berries *blue.* 1–3 ft. (30–90 cm). **Where found:** Rocky or peaty soils, swamps, bogs. Manitoba to Newfoundland and south to Minn., Mich., Penn., and New England. **Flowers:** April–July. **Fruit:** May–Aug.
Use: Fruit, jelly. The blueberrylike fruit are little used but excellent. Prepare them as you would blueberries. Look for them long before blueberries come into season.
                                  Summer (June, July)

CANADA BUFFALOBERRY   **Fruit**
Shepherdia canadensis
The only *opposite-leaved* northern shrub with mixed *silver and rusty scales* on twigs and leaf undersides. Leaves deciduous, elliptic to oval, dark green above; 1–1½ in. (2.5–3.8 cm) long. Small leaves may be in leaf axils. Fruit berrylike, orange or reddish, disagreeable tasting; clustered in leaf axils. 3–10 ft. (0.9–3 m). The more western SILVER BUFFALOBERRY, *S. argentea* (not shown), is somewhat thorny with silvery leaves and bright red fruit; the fruit are pleasantly tart and make fine jelly. **Where found:** Rocky and sandy soil, shores, streambanks. Canada south to Wisc., Maine. **Fruit:** July–Sept.
Use: Cold drink, jelly, emergency food. Although the berries contain a bitter substance called saponin, certain Indian tribes prized a salmon-colored drink made by frothing the berries in water and adding sugar. At least one author has suggested that the berries would make a good jelly. I found them unattractive and do not recommend them for general use. However, the berries are locally abundant and should be considered in emergencies.   Late Summer

OPPOSITE SIMPLE

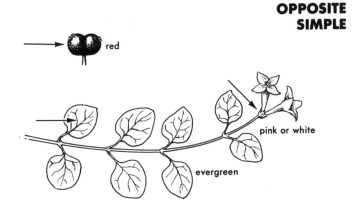

red

pink or white

evergreen

above: PARTRIDGEBERRY

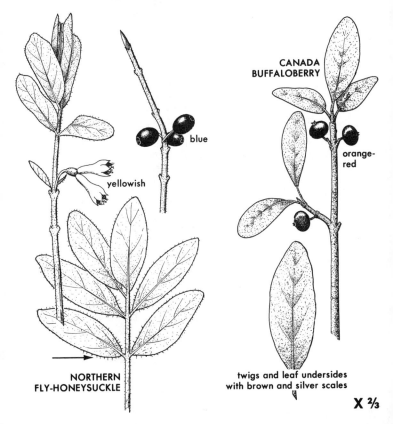

blue

yellowish

NORTHERN
FLY-HONEYSUCKLE

CANADA
BUFFALOBERRY

orange-
red

twigs and leaf undersides
with brown and silver scales

X ⅔

TREES WITH OPPOSITE
FAN-LOBED LEAVES

MAPLES   Sap
Acer spp.

Widespread, familiar trees with dry, double-winged fruit, "keys." One or more species found throughout; 2 shown. The leaves of some *Viburnums* (p. 178) are similar to those of the maples, but are either downy-hairy or have tiny glands on the stalks.

Use: Syrup, sugar, water. To obtain sap, bore a ½ in. (1.3 cm) hole 2–3 in. (5–7.5 cm) into the trunk of a tree, slanting upwards. Insert a spile, or spigot, into the hole, sharp end first. Commercial spiles are available in some hardware stores, or excellent ones can be made from Common Elderberry (p. 172): cut a 5 in. (12.5 cm) section from an elderberry branch, punch out the soft pith, make a long diagonal cut to form the sharpened end, and add a notch at the other end to hold a pail. Two spiles can be put into average-sized trees. Boil the sap slowly to make syrup—as much as 30–40 gal. of sap is needed to produce 1 gal. of syrup; further boiling produces sugar. As the syrup gets thicker, be sure not to let it boil over or burn. Syrup is ready to pour off when a candy thermometer reads 7°F (4°C) higher— 219°F (104°C) at sea level—than the temperature at which the sap first boiled; 22°F (12°C) higher for sugar—234°F (112°C) at sea level. The watery sap is quite pure and can be used as drinking or cooking water in areas where the water supply may be contaminated. Trees should be tapped in early spring (late Jan.–early April) when things begin to thaw but before the leaves appear; warm, sunny days after cold nights are the best times to tap. Although the sugar content varies from species to species, and even tree to tree, all maples produce excellent syrup.   LATE WINTER–EARLY SPRING

RED MAPLE, *A. rubrum.* A mid-sized tree. Young bark smooth, gray; older bark scaly, darker. Leaves 2–8 in. (5–20 cm) long; 3- to 5-lobed with fairly *shallow* notches between lobes; terminal lobe *broad* at base. Leaves *whitened beneath.* Twigs, buds, and fruit reddish. Height 20–40 ft. (6–12 m); diameter 1–2 ft. (30–60 cm). **Where found:** Wet woods. Se. Manitoba to Newfoundland, south to e. Texas and Fla. **Fruit:** May–July.

SUGAR MAPLE, *A. saccharum.* A tall tree with bark on older trees dark brown and deeply furrowed. Leaves usually *5-lobed;* hairless above, *light green beneath;* 2–10 in. (5–25 cm) long. Margins of leaves *not* curling down. Twigs glossy, reddish brown, with slender buds. Height 40–60 ft. (12–18 m); diameter 1–3 ft. (30–90 cm). **Where found:** Rich moist soil; upland forests. Sw. Manitoba to Newfoundland, south to e. Texas, Ark., Miss., Ala., and n. Ga. **Fruit:** June–Sept.

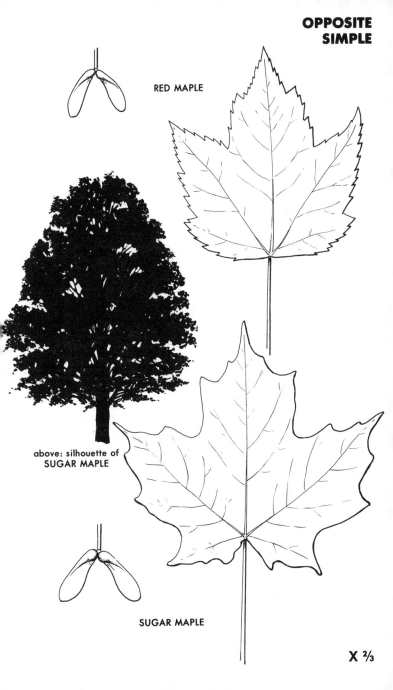

RED MAPLE

above: silhouette of
SUGAR MAPLE

SUGAR MAPLE

X ⅔

VIBURNUMS

WILD-RAISINS   **Fruit**
Viburnum spp.   **Color pl. 5 (Northern Wild-Raisin)**
Shrubs or small trees with flat-topped clusters of usually small, white flowers (see sketch) followed by small, fleshy, sweet tasting, black or blue-black berries containing *solitary large flat* seeds; flower and fruit clusters mostly 3–5 in. (7.5–12.5 cm) across. Berries often held into winter. Leaves egg-shaped or elliptic, usually fine-toothed. Several species in our area; 2 shown. **Note:** A similar group of *Viburnums,* Arrowheads (not shown), have coarsely-toothed leaves and bitter fruit.
Use: Nibble, cooked fruit, jelly. When fully ripe, the thin pulp is a pleasant raisin- or date-like nibble. Excellent cooked, the seeds removed, and mixed with one of the tart fruits in fruit stews, fruit sauces, and jellies.
                                              LATE SUMMER–FALL (WINTER)
NANNYBERRY, *V. lentago.* A tall shrub. Leaves smooth, abruptly long-pointed; 2–5 in. (5–12.5 cm) long. Leafstalks *winged.* Flowers small. Fruit blue-black; on slender, reddish, often drooping stalks. 10–20 ft. (3–6 m). **Where found:** Moist woods. Manitoba to New England, south to ne. Mo., N.J.; in mts. to Ga. **Flowers:** May–June. **Fruit:** Aug.-Sept.
HOBBLEBUSH, *V. alnifolium.* A straggling shrub whose branches often droop to the ground. Leaves heart-shaped, 4–8 in. (10–20 cm) long. Buds, twigs, and leaf undersides *rusty-hairy. Outer flowers large, showy.* Ripe fruit blue-black. To 10 ft. (3 m). **Where found:** Cool moist woods. S. Canada, n. U.S.; in mts. to Ga. **Flowers:** May–June. **Fruit:** Aug.-Oct.

HIGHBUSH-CRANBERRY   **Fruit**
Viburnum trilobum   **Color pl. 12**
A tall shrub similar to the wild-raisins (above), but leaves strongly 3-lobed and coarser toothed. Leaves 2–4 in. (5–10 cm) long. Lobes of leaves usually long-pointed; leafstalks with tiny dome-shaped glands at base of leaf blades. Flower clusters similar to those of Hobblebush (above). Ripe fruit tart, juicy, bright red. Height to 17 ft. (5.1 m). **Where found:** Cool woods, thickets, shores, rocky slopes. Canada, n. U.S. **Flowers:** May–July. **Fruit:** Sept.–Oct. (often remaining through winter).
Use: Cooked fruit, cold drink, jelly. Although not related to cranberries (p. 222), this fruit is often used in their place. Normally too tart to be eaten raw, the berries are excellent when cooked with some lemon peels, the seeds removed, and sugar added; the result is very like cranberry sauce. The sweetened juice can be diluted to make drinks or made into jelly with the addition of pectin. Rich in vitamin C. **Note:** A European ornamental occasionally escaped from cultivation, *V. opulus,* is almost a double for Highbush-cranberry, but with bitter fruit.
                                                   LATE SUMMER–FALL

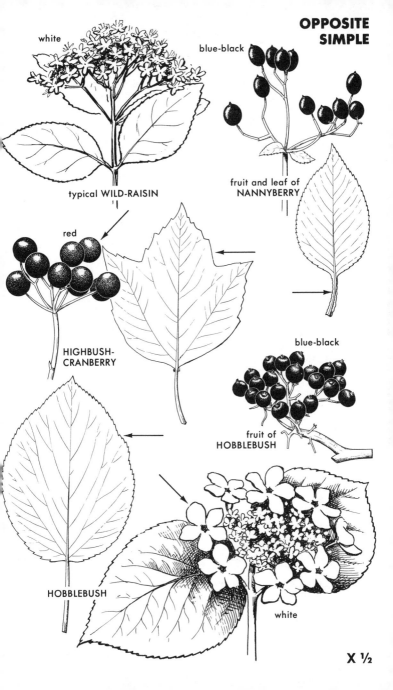

white

typical WILD-RAISIN

blue-black

fruit and leaf of NANNYBERRY

red

HIGHBUSH-CRANBERRY

blue-black

fruit of HOBBLEBUSH

HOBBLEBUSH

white

X ½

MISCELLANEOUS

SHRUBBY CINQUEFOIL   **Leaves**
Potentilla fruticosa   **Color pl. 6**
A low bushy shrub, often with loose, shreddy bark. Leaves 5-
(*to 7-*) *part;* leaflets less than 1 in. (2.5 cm) long, toothless, often
silky-whitish beneath. Flowers yellow, 5-petaled. Up to 3 ft.
(90 cm). **Where found:** Wet or dry open soil; meadows,
shores. Canada south to n. Iowa, n. N.J. **Flowers:** June–Oct.
Use: Tea. Steep the dried leaves for 5–10 min.   SUMMER

VIRGINIA CREEPER   **Poisonous**
Parthenocissus quinquefolia
A common climbing vine. Leaves long-stalked, palm-shaped,
with 5 (rarely 3 or 7) leaflets. Tendrils long, slender, *disk-tip-
ped.* Berries small, blue; in loose clusters. **Where found:**
Woods, thickets. Most of our area. **Fruit:** Aug.–Feb.
Warning: Berries potentially fatal if eaten in quantity.

KENTUCKY COFFEE-TREE   **Roasted Seeds (only)**
Gymnocladus dioica
A large tree. Leaves *huge, twice-compound;* 1–3 ft. (30–90 cm)
long. Leaflets numerous, toothless, oval. Branches and trunk
thornless. Bark is dark, furrowed. Flowers whitish, clustered.
Fruit purplish-brown pods 4–10 in. (10–25 cm) long, with 6–8
flattened seeds embedded in sticky pulp. Height 40–80 ft. (12–
24 m); diameter 1–2 ft. (30–60 cm). **Where found:** Rich soil;
woods. Mainly west of Appalachians, se. S. Dak. to cen. N.Y.,
south to w. Okla., Tenn., Va.; often cultivated east of Appalachians. **Flowers:** May–June. **Fruit:** Sept.–winter.
Use: Coffee. The seeds can be roasted and ground into a
caffeine-free coffee substitute. **Warning:** Use only the roasted
seeds; the fresh seeds and surrounding pulp are poisonous. Do
not mistake the pods for the much longer ones of Honey Locust, p. 184.   FALL–EARLY SPRING

WISTERIA   **Flowers (only)**
Wisteria spp.
Sturdy, twining vines with smooth bark, feather-compound
leaves, and *toothless* leaflets. Flowers in *showy, drooping clusters;* lilac to purple (or white), pealike. Seedpods somewhat
flattened, knobby. Several species; AMERICAN WISTERIA,
W. frutescens, shown. **Where found:** Floodplains, streambanks, thickets. One species or another from Mo. to se. Va.,
south to e. Texas and Fla.; cultivated elsewhere. **Flowers:**
April–June. **Fruit:** Sept.–Nov.
Use: Fritters. The flowers can be dipped into batter and fried
in oil to make excellent fritters. **Warning:** The seeds are quite
poisonous; 2 are enough to harm a child.   SPRING

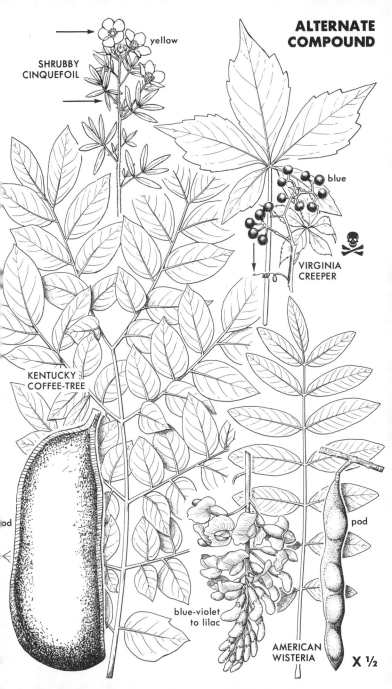

ALTERNATE COMPOUND

SHRUBBY CINQUEFOIL

yellow

VIRGINIA CREEPER

blue

KENTUCKY COFFEE-TREE

pod

blue-violet to lilac

pod

AMERICAN WISTERIA

X ½

3 LEAFLETS; THORNLESS VINES, SHRUBS

POISON IVY   **Poisonous**
Rhus radicans
A trailing or climbing vine, or erect shrub. Leaves on long stalks; 4–14 in. (10–35 cm) long. Leaflets highly variable; hairless or slightly hairy, glossy or dull, toothless or saw-toothed and variously lobed. End leaflet pointed, longer-stalked than side pair. Climbing stems (only) with short aerial rootlets along sides and covered with dark fibers when old. Berries small, hard, white; in drooping clusters. Height 2–7 ft. (0.6–2.1 m) when not climbing. POISON OAK, *R. toxicodendron,* is similar but always *erect;* leaflets *hairy,* often lobed like an oak leaf. **Where found:** Woods, thickets. S. Canada and south to Texas and Fla. **Fruit:** Aug.–Nov.
Warning: Do not touch. Contact with any part can result in severe dermatitis; washing with soap and water or jewelweed juice (p. 78, p. 92) within a few hours often prevents rashes.

SCOTCH BROOM   **Flowerbuds and young pods,**
Cytisus scoparius   **seeds**
A dense, *stiff-branched, deciduous or evergreen* shrub. Stems angled, dark green; leaves up to ½ in. (1.3 cm) long, usually 3-part but also lance-shaped. Flowers yellow, pealike, showy. Fruit flattened pods. Height 3–5 ft. (0.9–1.5 m). **Where found:** Dry sandy soil, roadsides, thickets. W. New York, sw. Maine, Nova Scotia, south to W. Va. and Ga., especially near coast.
Flowers: May–June. **Fruit:** July–Sept.
Use: Pickle, coffee. The unopened *flowerbuds* and *young seedpods* can be pickled, but only after they have been soaked overnight in a strong solution of vinegar and salt and then thoroughly rinsed; weigh them down so they will not float. The roasted *seeds* can be used to make a coffeelike beverage.
Warning: The raw seedpods are mildly poisonous.

SUMMER

KUDZU   **Roots**
Pueraria lobata
A long — 30–100 ft. (10–30 m) — trailing or high-climbing, semi-woody vine with downy young stems; often completely covers trees upon which it climbs. Leaves large; leaflets broadly oval, often 2- or 3-lobed, downy beneath. Flowers pealike, purple-violet or tinged with red, with a conspicuous yellow patch on upper petal; *flowers smell of grapes.* Root large, forking. **Where found:** Thickets, borders of woods. Tenn. and Va., south through southeastern states. **Flowers:** July–Sept.
Use: Flour. Stripped of their tough outer covering, the fleshy inner cores of the large root-branches can be cut into pieces, crushed in cold water, and the starch allowed to settle. The starch can then be refined by repeated washings in cold water and dried to make a fine white flour.   ALL YEAR

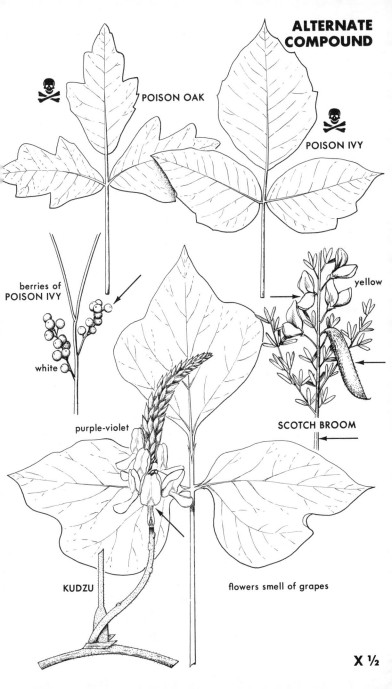

ALTERNATE COMPOUND

POISON OAK

POISON IVY

berries of
POISON IVY

yellow

white

purple-violet

SCOTCH BROOM

KUDZU

flowers smell of grapes

X ½

TREES AND SHRUBS WITH THORNS

ROSES   **Petals, fruit (hips)**
Rosa spp.   **Color pl. 1 (Wrinkled Rose)**
Familiar prickly or bristly shrubs. Leaves once-compound, usually with 5–11 leaflets. Note leafy stipules at bases of leaf-stalks. Flowers pink or white, 5-petaled, showy. Fruit reddish, with 5 conspicuous calyx lobes, mealy-textured pulp, and numerous seeds. Many species (see p. 106); typical form shown. **Where found:** Fields, fencerows, swamps. Throughout.
Use: Jam, tea, candy. See p. 106.   SUMMER (petals);
LATE SUMMER–EARLY FALL (fruit)

BRAMBLES   **Young shoots (blackberries); leaves, fruit**
Rubus spp.   **Color pl. 11 (Common Blackberry)**
A complex group of familiar rambling or arching, non-climbing, thorny or bristly shrubs. Leaves generally 2–10 in. (5–25 cm) long, with 3–7 leaflets. Stems usually green or red. Showy white 5-petaled flowers are followed by juicy red or black fruit. To 6 ft. (1.8 m). 3 major subgroups: (1) Raspberries, with arching, usually *white-powdered, round stems*. Red fruit form hollow shells when separated from stalks. (2) Dewberries, *flattened* shrubs less than 1 ft. (30 cm) high with round or angular *trailing* stems and blackberrylike fruit. (3) Blackberries, with *angular,* arching stems. **Where found:** Sunny thickets. Throughout. **Flowers:** April–July. **Fruit:** June–Sept.
Use: Fruit, jelly, cold drink, tea, salad. In addition to the familiar uses for the *fruit,* the dried *leaves* can be used to make tea and tender *blackberry shoots* can be added fresh to salads.
SPRING (blackberry shoots); SUMMER (leaves, fruit)

HONEY LOCUST   **Pulp of unripe seedpods**
Gleditsia triacanthos
Taller than Black Locust (below), with once- or twice-compound leaves. Leaflets narrow, may be vaguely toothed. Branches and trunk armed with *long, stout, branched thorns.* Pods 8–18 in. (20–45 cm), flattened, twisted. **Where found:** Rich woods, fields. Northern U.S. south. **Fruit:** Sept.–Feb.
Use: Nibble. Thin pulp of unripe pods sugary sweet. **Caution:** See pods of Kentucky Coffee-tree, p. 180.   SUMMER

BLACK LOCUST   **Flowers (only)**
Robinia pseudo-acacia   **Color pl. 11**
A mid-sized tree. Leaves once-compound, 6–12 in. (15–30 cm) long, with 6–20 oval leaflets. Twigs smooth, stout; with paired thorns at bases of leafstalks. Bark dark gray, deeply ridged. Flowers white, in *showy drooping clusters;* pealike, very fragrant. Fruit a flat pod. **Where found:** Dry woods, old fields. S. Canada south to e. Okla., Ga. **Flowers:** April–June.
Use: Fritters. The flower clusters make outstanding fritters.
Warning: Roots, bark, leaves, and seeds poisonous. SPRING

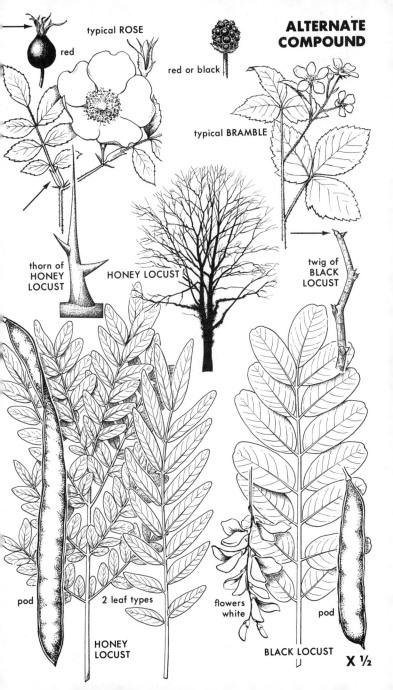

ALTERNATE COMPOUND

red

typical ROSE

red or black

typical BRAMBLE

thorn of HONEY LOCUST

HONEY LOCUST

twig of BLACK LOCUST

pod

2 leaf types

HONEY LOCUST

flowers white

pod

BLACK LOCUST

X ½

SUMACS

POISON SUMAC   **Poisonous**
Rhus vernix
A shrub or small tree. Leaves 6–12 in. (15–30 cm) long, with 7–13 pointed, *toothless* leaflets. Twigs and buds *hairless*. Bark smooth, dark, speckled with dark spots. Note the spreading or drooping clusters of small, *ivory-white* berries. Height 6–20 ft. (1.8–6 m). **Where found:** Partly wooded swamps. Se. Minn., s. Ontario, sw. Quebec, sw. Maine south to Texas and Fla. **Fruit:** Aug.–spring.
Warning: Do not touch. More virulent than Poison Ivy (p. 182). Contact with any part can cause severe dermatitis.

SUMACS   **Fruit**
Rhus spp.
Shrubs or small trees with large feather-compound leaves. Twigs stout, pithy, with *milky sap*. Note the dense clusters of small, dry, *hairy, red fruit* at the end of the branches; old clusters persist, making the winter silhouette distinctive. 3 species and silhouette of Staghorn Sumac shown.
Use: Cold drink. When ripe, the hard berries are covered with acidic red hairs. Collect the entire fruit cluster, rub gently to bruise the berries, and soak for 10–15 min. in *cold* water. Remove the cluster, and pour the pink juice through cheesecloth to strain out the hairs and any loose berries. Sweeten to taste and chill; tastes like pink lemonade. Gather the clusters before heavy rains wash out most of the acid.   SUMMER
DWARF OR WINGED SUMAC, *R. copallina.* Leaves 6–14 in. (15–35 cm) long with 11–23 glossy, toothless leaflets; midrib between leaflets with conspicuous *wings.* Twigs and leafstalks *velvety.* Height 4–10 ft. (1.2–3 m). **Where found:** Upland fields and openings. E. Kans., cen. Wisc., Mich., se. N.Y., s. Maine, south to e. Texas and Fla. **Fruit:** Aug.–Oct.
SMOOTH SUMAC, *R. glabra.* **Color pl. 11.** Resembles Staghorn Sumac (below) but with *hairless* twigs and leafstalks. Hybrids between the two occur. Twigs slightly flattened. 4–15 ft. (1.2–4.5 m). **Where found:** Old fields, fencerows. S. Canada, south to Texas and nw. Fla. **Fruit:** June–Oct.
STAGHORN SUMAC, *R. typhina.* Leaves 1–2 ft. (30–60 cm) long with 11–31 toothed leaflets. Twigs round; twigs and leafstalks very hairy. Bark smooth, dark, with numerous raised crosslike streaks. Up to 30 ft. (9 m). **Where found:** Upland old fields and openings. Minn., s. Ontario, e. Quebec, Nova Scotia, south to ne. Iowa, Ill., cen. Tenn., n. Ga., Md. **Fruit:** June–Sept.

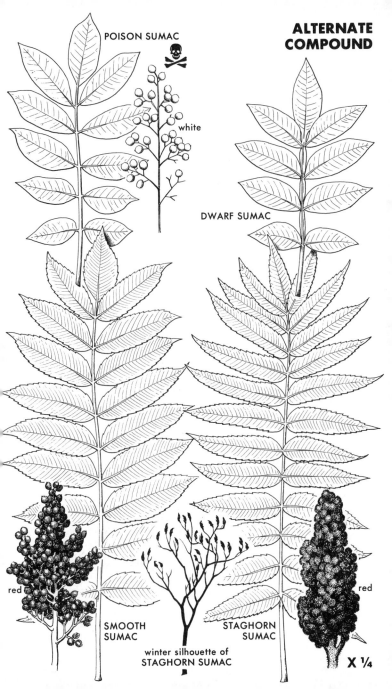

ALTERNATE COMPOUND

POISON SUMAC

white

DWARF SUMAC

SMOOTH SUMAC

winter silhouette of STAGHORN SUMAC

STAGHORN SUMAC

red

red

X ¼

MOUNTAIN-ASHES AND WALNUTS

MOUNTAIN-ASHES   **Fruit**
Pyrus spp.
Shrubs or small trees with once-compound leaves and showy
flat-topped clusters of white flowers and reddish fruit. Leaves
with 11–17 narrow, toothed leaflets. Bark gray, smoothish.
Use: Fresh or cooked fruit, jelly. Although bitter and unpleasant tasting when not fully ripe, the mealy-textured fruit become pleasantly sour after repeated freezings. They can be
eaten raw, but are better cooked and sweetened. Rich in pectin.
                                        FALL (after frosts)–EARLY SPRING
AMERICAN MOUNTAIN-ASH, *P. americana* (leaf shown).
Leaves 6–9 in. (15–22.5 cm); leaflets *3–5 times longer than
broad.* Flowers and fruit $\frac{1}{4}$ in. (0.6 cm). Up to 40 ft. (12 m).
Where found: Woods, openings. S. Canada and n. U.S.; south
in mts. to n. Ga. **Fruit:** Aug.–spring.
NORTHERN MOUNTAIN-ASH, *P. decora.* **Color pl. 15.**
Similar, but leaflets broader; flowers and fruit slightly larger,
showier. **Where found:** Cool moist woods, openings. Canada,
n. U.S. **Fruit:** Sept.–spring.

WALNUTS   **Nuts, sap**
Juglans spp.
Medium to tall trees with large once-compound leaves that
have 7–17 narrow, toothed leaflets. Fruit with a distinctive
thick, green, fleshy husk that is in *one closely bound unit*
around a rough-surfaced blackish-brown nutshell. Bark rough
and furrowed.
Use: Nuts, candy, flour, oil, syrup, sugar, water. Although the
nutshells are often difficult to open, the *nuts* are sweet and
delicious; they can be eaten out of hand, dipped in sugar syrup
and eaten as candy, ground into a meal-like flour, or crushed
and boiled to separate out an excellent vegetable oil. Although
walnuts are becoming increasingly scarce, a single tree will
produce a large supply of nuts; gather the nuts when they fall
to the ground. The *sap* can be used in the same way as maple
sap (p. 176).   EARLY SPRING (sap); FALL (nuts)
BUTTERNUT, *J. cinerea.* Small. Leaves similar but end
leaflet usually present. Wider bark ridges smooth-topped,
gray-shiny. Fruit oblong and sticky. Height to 80 ft. (24 m);
diameter to 2 ft. (60 cm). **Where found:** Rich soil, deciduous
woods. Se. Minn., s. Ontario, and w. New Brunswick, south to
n. Ark., n. Miss., w. Ga., and w. S.C. **Fruit:** Oct.–Nov.
BLACK WALNUT, *J. nigra.* The mildly spicy-scented leaves
often *lack* an end leaflet; leaves 1–2 ft. (30–60 cm) long. Bark is
dark, deeply furrowed; ridges not shiny. Fruit large, spherical.
Height to 100 ft. (30 m); diameter to 4 ft. (1.2 m). **Where
found:** Rich soil, deciduous woods. S. Minn., s. Ontario,
w. Mass., south to e. Texas, nw. Fla. **Fruit:** Oct.–Nov.

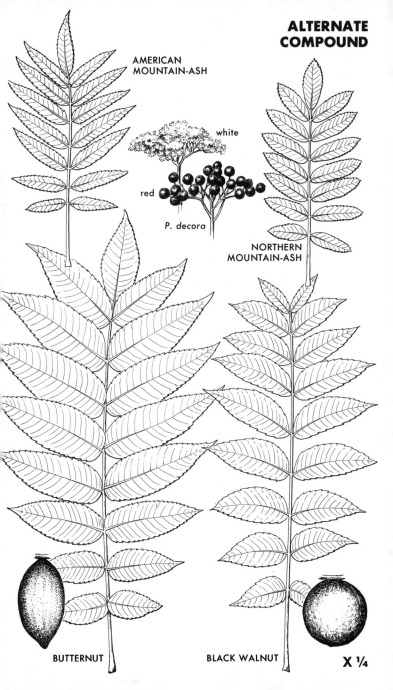

ALTERNATE COMPOUND

AMERICAN MOUNTAIN-ASH

white

red

P. decora

NORTHERN MOUNTAIN-ASH

BUTTERNUT

BLACK WALNUT

X ¼

HICKORIES   Nuts, sap
Carya spp.

Trees that are somewhat similar to the walnuts (p. 188) but leaves usually have 9 or fewer leaflets (Pecan has 9–17 leaflets); the final 3 leaflets are often larger than the rest, with the terminal leaflet largest of all. The fruit are smaller than those of the walnuts and have *husks that split into 4 sections* when mature, rather than remaining as a single unit. A number of species in our area; some sweet and edible, others bitter and inedible; 3 of the best shown.

Use: Nuts, candy, flour, oil, syrup, sugar, water. Gather the *nuts* when they fall to the ground, usually after most of the leaves have fallen. Use the nuts as you would walnuts (p. 188) and the *sap* as you would maple sap (p. 176).

EARLY SPRING (sap); FALL (nuts)

SHAGBARK HICKORY, *C. ovata.* Tall. Leaves 8–14 in. (20–35 cm) long, with 5–7 (usually 5) leaflets. Twigs stout, red-brown, hairless or nearly so. Bark light-colored, *very shaggy.* Nuts egg-shaped, with *thick-walled husks.* Up to 90 ft. (27 m). **Where found:** Rich soil, river bottoms, upland slopes. Se. Neb., se. Minn., cen. Mich., s. Ontario, sw. Quebec, and sw. Maine, south to e. Texas and nw. Fla.

PECAN, *C. illinoensis.* Tall. Leaves 12–30 in. (30–75 cm) long, divided into 9–17 leaflets; leaflets narrow, glossy, asymmetrical, often recurved at tip. Bark is dark, with many vertical ridges. Nuts surrounded by smooth, *oblong, thin-shelled husks.* Height to 120 ft. (36 m); diameter to 4 ft. (1.2 m). **Where found:** Rich moist soil, bottomland woods. Iowa to Ind., south to cen. Texas and Ala.

MOCKERNUT HICKORY, *C. tomentosa.* Medium-sized to tall. Leaves 8–15 in. (20–37.5 cm) long, with 7–9 leaflets; leaf undersides and twigs *matted-woolly.* Bark tight and deeply furrowed. Tough nuts surrounded by a ball- or egg-shaped, *thick-walled husk* that does *not* split to the base. Up to 80 ft. (24 m). **Where found:** Dry to moist upland woods. Neb., s. Mich., s. Ontario, N.Y., and s. N.H., south to e. Texas, n. Fla.

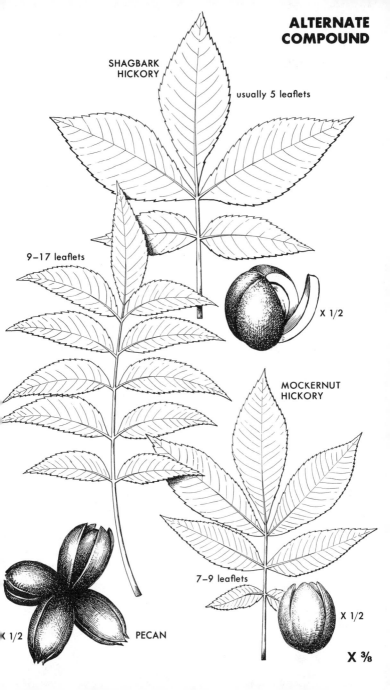

ALTERNATE COMPOUND

SHAGBARK HICKORY

usually 5 leaflets

9–17 leaflets

X 1/2

MOCKERNUT HICKORY

7–9 leaflets

X 1/2

X 1/2 PECAN

X ³⁄₈

MISCELLANEOUS SHRUBS; SMALL TREES

REDBUD   **Flowerbuds, flowers, young pods**
Cercis canadensis
A small tree that becomes cloudy with small, pinkish, pealike flowers before the leaves appear. See p. 122.

    EARLY SPRING (buds, flowers); EARLY SUMMER (pods)

YAUPON, CASSINA   **Leaves**
Ilex vomitoria
A large, shrubby, *evergreen* holly. Leaves up to 2 in. (5 cm) long, leathery; glossy deep green above and pale green below; scalloped-toothed. Bark whitish-gray. Note the conspicuous clusters of small berrylike red fruit. Up to 20 ft. (6 m). **Where found:** Moist sandy woods, clearings, thickets. Coastal plain; se. Va. to cen. Fla., west to s. Texas, and north in Mississippi Valley to se. Okla., n. Ark. **Fruit:** Sept.–March.
Use: Tea. Excellent dried in an oven until almost black, then steeped in hot water; contains caffeine. Other hollies, *Ilex* spp., can also be used but contain no caffeine. **Warning:** Berries can cause vomiting and diarrhea.   ALL YEAR

COMMON BARBERRY   **Fruit**
Berberis vulgaris   **Color pl. 11**
A thorny shrub with slender arching branches. Leaves alternate or in whorled clusters; wedge-shaped, closely bristle-toothed. *Thorns usually with 2 side branches* as long as central spike. Inner bark and wood yellow. Small yellow flowers are followed by pendent clusters of very tart, juicy red fruit. Up to 10 ft. (3 m). AMERICAN BARBERRY, *B. canadensis* (not shown), has fewer teeth and ball-like fruit; dry open woods from se. Mo. and w. Va. to Ga. **Where found:** Old fields, thickets. S. Canada south to Mo., Del. **Fruit:** Aug.–spring.
Use: Jelly, cooked fruit, cold drink. Barberries make an outstanding jelly or cooked fruit. The diluted and sweetened juice makes a fine cold drink. Rich in pectin.   FALL

PAWPAW   **Fruit**
Asimina triloba
Leaves toothless, 6–12 in. (15–30 cm) long; dark green above, lighter beneath. Flowers precede leaves; 6-petaled, purple. *Fruit suggest stubby bananas;* green, then brown when ripe. 6–20 ft. (1.8–6 m). **Where found:** Rich soil; streambanks, woods. S. Iowa to N.J., south to e. Texas, Fla. **Fruit:** Aug.–Oct.
Use: Fruit. The sweet, creamy-textured, yellow pulp is delicious raw or cooked. Gather while still green, and set aside for a few days to ripen to dark brown.   FALL

NEW JERSEY TEA   **Leaves**
Ceanothus americanus   **Color pl. 13**
A low shrub; woody only near its base. See p. 18.   SUMMER

192

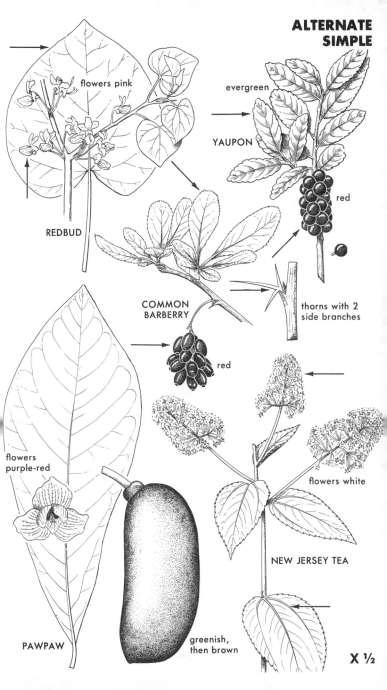

**ALTERNATE
SIMPLE**

flowers pink

evergreen

YAUPON

red

REDBUD

COMMON
BARBERRY

thorns with 2
side branches

red

flowers white

flowers
purple-red

NEW JERSEY TEA

PAWPAW

greenish,
then brown

X ½

MISCELLANEOUS TREES

PERSIMMON   Fruit, leaves
Diospyros virginiana
Bark nearly black, characteristically in small *rectangular blocks.* Leaves stiff, oval, toothless; dark-green glossy above, paler beneath; 2–6 in. (5–15 cm) long. *Buds 2-scaled, very dark.* Fruit about 1½ in. (3.8 cm) across, with 6 flattened seeds in center; orange to reddish purple when ripe. Height to 60 ft. (18 m); diameter to 2 ft. (60 cm). **Where found:** Old fields, dry deciduous woods. E. Kans. to Conn., and south to e. Texas and Fla. **Fruit:** Aug.–Oct. (or later).
Use: Fresh fruit, jam, pudding, nut bread, tea. Persimmon *fruit* are unpleasantly astringent when green, but richly sweet when fully ripe; collect them when they are soft and gooey, and remove the seeds. To gather fruit in quantity, spread a sheet under a tree and shake the branches; separate out the firm unripe persimmons. The dried *leaves* can be made into an excellent, full-bodied tea rich in vitamin C.
                    SUMMER (leaves); LATE FALL–MIDWINTER (fruit)

AMERICAN HACKBERRY   Fruit
Celtis occidentalis
A small to large tree. Leaves 3–6 in. (7.5–15 cm) long, singly toothed, rough-hairy (or smooth) above; leaf tips long, often slightly curved; leaf bases uneven. *Older bark corky or warty,* gray to brown. Fruit dark red to purple, borne singly on slender stems in leaf axils; thin, sweet, dry pulp surrounds a solitary seed. Height 20–70 ft. (6–21 m); diameter 1–3 ft. (30–90 cm). SUGARBERRY, *C. laevigata,* found in southern lowlands is taller, with toothless leaves and yellow to orange fruit. **Where found:** Rich, moist soil; woods. Idaho to sw. Quebec, south to Okla., Ark., and n. Fla.; more common southward. **Fruit:** Oct.–Nov. (or later).
Use: Nibble. The sugary pulp is delicious.
                                         FALL–EARLY WINTER

AMERICAN BASSWOOD   Leaf buds, flowers
Tilia americana
A tall tree. Leaves 5–10 in. (12.5–25 cm) long, *finely-toothed,* heart-shaped with *uneven* bases, *hairless.* Bark dark, shallowly grooved; smooth gray on upper parts. Clusters of yellowish flowers dangle from long, conspicuously *winged* stalks. Height 60–80 ft. (18–24 m); diameter to 3 ft. (90 cm). **Where found:** Moist woods. S. Canada south to Texas and Ala. **Flowers:** June–Aug.
Use: Salad (nibble), tea. The unopened *leaf buds* are slightly mucilaginous and make an excellent trailside nibble. The dried *flowers* make a mild, pleasant tea. Store excess flowers in sealed containers.   EARLY SPRING (buds); SUMMER (flowers)

194

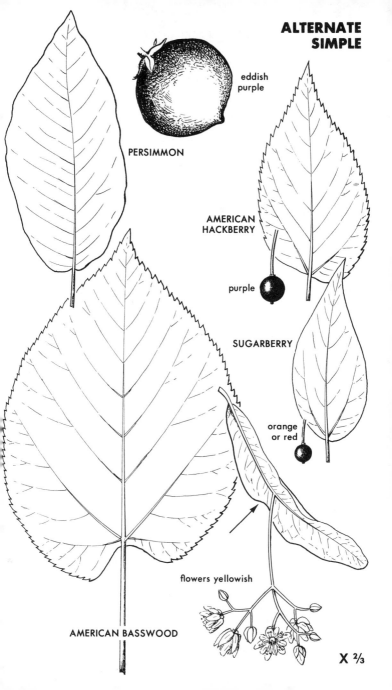

ALTERNATE SIMPLE

eddish purple

PERSIMMON

AMERICAN HACKBERRY

purple

SUGARBERRY

orange or red

flowers yellowish

AMERICAN BASSWOOD

X ⅔

PRICKLY GREEN-STEMMED VINES
WITH TENDRILS

GREENBRIERS, CATBRIERS   **Young shoots and**
Smilax spp.   **leaves, rootstock**

Green-stemmed, mostly prickly or thorny vines climbing by
tendrils that originate in the leaf axils. Leaves parallel-veined;
stems with no central pith. Leaves on some southern species
remain through winter. Flowers small and greenish; fruit are
small berries with 1 or several large seeds.

Use: Asparagus, salad, cooked green; jelly, flour, cold drink.
The young *shoots* are excellent cooked like asparagus and
served with butter. The *young shoots, leaves, and tendrils* are
also quite good prepared like spinach or added fresh to salads.
Use the young shoots and leaves for as long as they remain
tender and juicy — often well into summer. The *rootstocks* of
many species yield a gelatin substitute. Crush them thoroughly, and prepare them using the washing process described
for cattails (p. 158). The dry red powder that is produced
makes a mild jelly when boiled with water (1 tbs. powder per
cup of water). Also, it can be mixed half and half with wheat
flour, added to stews as a thickening agent, or diluted and
sweetened for use as a cold drink. Of the 3 species shown, the
tuberous roots of Bullbrier (*S. bona-nox*) are by far the best to
use.   SPRING–SUMMER (shoots, leaves);
ALL YEAR (rootstock)

COMMON GREENBRIER, CATBRIER, *S. rotundifolia.*
Leaves *broadly rounded* or heart-shaped with smooth margins.
Stems rounded or angled, well-armed with strong prickles.
Rootstock long and thin. Fruit blue-black, dusted with powder,
usually 2-seeded. **Where found:** Woods, thickets. Okla.,
se. Mo., Ill., s. Mich., s. Ontario, N.Y., s. N.H., s. Maine, and
Nova Scotia, south to Texas and Fla. **Fruit:** Sept.–spring.

LAUREL GREENBRIER, *S. laurifolia.* A vigorous, high-
climbing *evergreen* vine with elliptic, *leathery* leaves. Stems
rounded, may have prickles near base and on young shoots.
Fruit black. **Where found:** Swamps, bottomlands. Tenn. and
N.J., south to Texas and Fla. **Fruit:** Aug.–Sept.

BULLBRIER GREENBRIER, *S. bona-nox.* Highly variable,
with at least 4 forms. Typical form is shrubby, not a high
climber. Leaves leathery, bristly-edged, *triangular to fiddle-
shaped, often mottled* with white; 2–5 in. (5–12.5 cm) long.
Other varieties may be high-climbing, have slender leaves with
heart-shaped bases, or may lack mottling and fringed leaf-
edges. All varieties have 4-sided stems covered with stiff
prickles. Rootstock large, thick, and knobby. Fruit black,
1-seeded. **Where found:** Woods, thickets. Kans., Mo., s. Ill.,
se. Ind., Ky., Md., Del., and se. Mass., south to Texas and Fla.
Fruit: Oct.–spring.

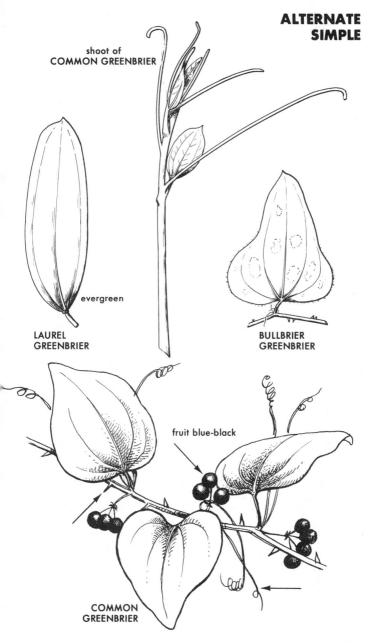

shoot of
COMMON GREENBRIER

evergreen

LAUREL
GREENBRIER

BULLBRIER
GREENBRIER

fruit blue-black

COMMON
GREENBRIER

X ⅔

THORNLESS CLIMBING OR TWINING VINES

WILD GRAPES   Young leaves, fruit
Vitis spp.
Thornless, high-climbing vines with *forked tendrils*. Stems dark, mostly shreddy-barked. Branchlets usually have a *brownish pith* which is interrupted by partitions at the leaf nodes. Leaves large, coarsely toothed, often lobed; heart-shaped, especially at base. Flowers greenish. Fruit fleshy, each with 1–4 *pear-shaped* seeds; purple, black, or amber. Typical form shown. **Where found:** Thickets, edges of woods. One or more species throughout most of our area. **Fruit:** Aug.–Oct. **Use:** Fresh fruit, jelly, cold drink, cooked green. Wild grapes can be used in any recipe that calls for grapes; generally, they need more sweetening than cultivated grapes. Before fully ripe, the *fruit* are an excellent source of pectin. The *young leaves* can be boiled for 10–15 min. and served with butter; or they can be lightly boiled and used to wrap rice or meat for baking. Collect the leaves in early summer while still tender but fully unfurled.   EARLY SUMMER (leaves); FALL (fruit)

CANADA MOONSEED   Poisonous
Menispermum canadense
A climbing vine with green twining stems. Leaves large — 5–10 in. (12.5–25 cm) — variable, sometimes nearly round with a pointed tip, but usually with 3–7 shallow lobes. Leaf bases *not attached* to leafstalks. Drooping clusters of white flowers develop into soft grape-sized black fruit powdered with white. Each fruit has *a single, flattened, crescent-shaped seed.* **Where found:** Moist woods, thickets, streambanks. Se. Manitoba, w. Quebec, w. New England, south to Okla., Ark., Ala., and Ga. **Flowers:** May–July. **Fruit:** Sept.–Oct. **Warning:** Although this vine strongly suggests a wild grape, it lacks tendrils and has only a single crescent-shaped seed. The bitter, unpalatable fruit are potentially fatal if eaten in large enough quantities.

NIGHTSHADE, BITTERSWEET   Poisonous
Solanum dulcamara
A weak woody-stemmed vine trailing over bushes. Flowers with 5 recurved violet petals and *beaklike* yellow anthers. Leaves usually with *2 small lobes* at bases. Berries oval, in drooping clusters; green, then bright translucent red. 2–8 ft. (0.6–2.4 m). **Where found:** Moist thickets, clearings. Minn., Ontario, Nova Scotia south to Kans., Mo., and Ga. **Flowers:** May–Oct. **Fruit:** Aug.–May. **Warning:** The ruby-red berries are quite visible and attractive looking, but bitter and poisonous.

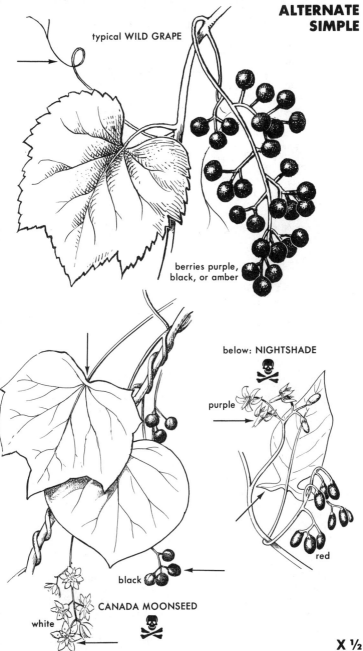

ALTERNATE SIMPLE

typical WILD GRAPE

berries purple, black, or amber

below: NIGHTSHADE

purple

red

black

CANADA MOONSEED

white

X ½

DOUBLE-TOOTHED LEAVES

(SWEET) BIRCHES   Sap, inner bark, twigs
Betula spp.
Medium-sized northern trees. Leaves finely double-toothed, somewhat egg-shaped with sharp tips and blunt bases. *Crushed twigs smell and taste of wintergreen.* 2 species.
Use: Syrup, sugar, water, flour, tea. Birch *sap* flows abundantly and can be processed like maple sap (p. 176) to make a sweet, molasseslike syrup; the flow is usually best in late March or April. The *inner bark* can be dried and ground into flour for emergency use. The *twigs* can be steeped in hot water to make tea. **Note:** The sap of all birches, *Betula* spp., is edible.
SPRING (sap, inner bark); ALL YEAR (twigs)
BLACK BIRCH, *B. lenta.* Young bark *dark and tight;* old bark broken into irregular plates. Height to 70 ft. (21 m); diameter to 3 ft. (90 cm). **Where found:** Mature forests. E. Ontario to sw. Maine, south to Ohio, Del.; in mts. to n. Ga.
YELLOW BIRCH, *B. lutea.* Taller. Bark *yellowish to silver-gray,* peeling in narrow curls. **Where found:** Damp forests. Canada south to ne. Iowa, n. Ind., Md., Del.; in mts. to n. Ga.

SLIPPERY ELM   Inner bark
Ulmus rubra
A small or medium-sized tree. Leaves 5–8 in. (12.5–20 cm) long; *rough and sandpapery* above, *hairy* beneath. Twigs *rough-hairy.* Older bark grayish; in flat-topped, nearly vertical ridges. Solitary seeds encased in round, waferlike, smooth-surfaced wings. Height to 60 ft. (18 m); diameter to 2 ft. (60 cm). **Where found:** Rich soil, woods. N. Dak. to New England, south to Texas and nw. Fla. **Fruit:** May–June.
Use: Tea, flour. The slimy inner bark makes a pleasantly wholesome tea when steeped in hot water for 15 min.; dried and ground, it becomes a nutritious flour.   SPRING

HAZELNUTS   Nuts
Corylus spp.
Tall shrubs. Leaves 2–5 in. (5–12.5 cm) long, coarsely *double-toothed,* with somewhat *heart-shaped* bases. Nuts encased in leafy husks. Height to 10 ft. (3 m). 2 species.
Use: Nuts, flour, candy. Roasted nuts excellent eaten as is, ground into flour, or candied.   LATE SUMMER–EARLY FALL

BEAKED HAZELNUT, *C. cornuta.* Twigs usually *smooth.* Husks *beaklike, bristly,* completely surrounding nuts. **Where found:** Thickets, edges of woods. S. Canada south to e. Kans. and Ga. **Fruit:** Aug.–Sept.
AMERICAN HAZELNUT, *C. americana.* Twigs and leaf-stalks *rough-hairy.* Husks *ragged-edged,* open at one end to expose nuts. **Where found:** Thickets. Saskatchewan to Maine, south to Okla., Mo., and Ga. **Fruit:** July–Sept.

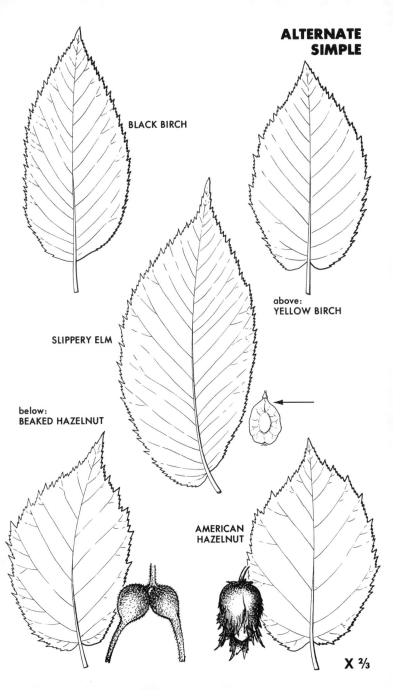

ALTERNATE SIMPLE

BLACK BIRCH

above:
YELLOW BIRCH

SLIPPERY ELM

below:
BEAKED HAZELNUT

AMERICAN HAZELNUT

X ⅔

BRISTLY-HUSKED NUTS;
COARSELY-TOOTHED LEAVES

AMERICAN BEECH   **Nuts**
Fagus grandifolia
Tall trees with distinctive *smooth gray bark* and coarsely-
toothed elliptic leaves. Leaves 1–5 in. (2.5–12.5 cm) long. Fruit
small triangular nuts enclosed in a bur-like husk with weak
spines; each husk contains 2–3 nuts. Height 60–80 ft.
(18–24 m); diameter 2–3 ft. (60–90 cm). **Where found:** Rich
moist soil; upland forests. E. Wisc., s. Ontario, Nova Scotia,
south to Texas and n. Fla. **Fruit:** Sept.–Oct.
Use: Nuts, flour, oil, coffee. The thin-shelled nuts have sweet
kernels that are delicious roasted and eaten whole, or ground
into flour. An outstanding vegetable oil can be squeezed from
the crushed kernels. The roasted kernels can be used as a coffee
substitute. For some reason usually only the trees in the n. U.S.
and Canada produce plentiful supplies of nuts; gather them
after they drop from the trees during the first frosty nights in
October.   FALL

CHESTNUT   **Nuts**
Castanea dentata
Because of the Chestnut blight, this species now occurs mostly
as sprouts from old stumps. Leaves large, spear-shaped,
coarsely toothed, *hairless*. Distinctive *stiff-prickly* husks contain *2 or 3* slightly flattened nuts. Height rarely exceeds 15 ft.
(4.5 m). **Where found:** Well-drained forests. S. Mich. to
s. Maine, south to ne. Miss. and Ga. **Fruit:** Sept.–Oct.
Use: Nuts, flour, candy. Nowadays few trees reach sufficient
maturity to produce more than a few nuts. If you do find
enough to make collection worthwhile, gather the nuts after the
first frost splits the husks open. Once the husks and bitter pith
are removed, the kernels are sweet and delicious.   FALL

EASTERN CHINQUAPIN   **Nuts**
Castanea pumila
Similar to Chestnut (above). Rarely tree-sized. Leaves *white-
hairy beneath;* 3–5 in. (7.5–12.5 cm) long. Husks and nuts
smaller; nuts occur *singly* in husks, not flattened. Height 10–15
ft. (3–4.5 m); diameter 1–2 in. (2.5–5 cm). **Where found:** Dry
upland woods, thickets. Ark., Tenn., e. Penn., and N.J., south
to e. Texas and Fla. **Fruit:** Sept.–Oct. OZARK CHINQUA-
PIN, *C. ozarkensis* (not shown), is similar but much taller;
Okla. and s. Mo., south to La. and Miss.
Use: Nuts, flour, candy. Roast the nuts and crack the shells to
remove the kernels. The kernels can be eaten as is, ground into
flour, or dipped in sugar syrup to make candy.   FALL

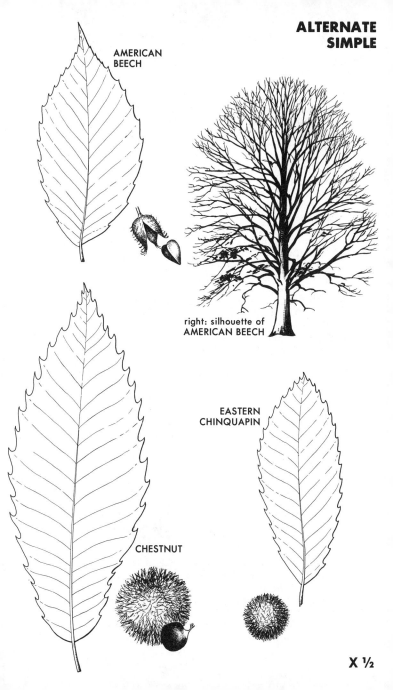

ALTERNATE SIMPLE

AMERICAN BEECH

right: silhouette of AMERICAN BEECH

EASTERN CHINQUAPIN

CHESTNUT

X ½

TREES WITH ACORNS: OAKS

OAKS   **Acorns**
Quercus spp.

A large, widespread group of familiar trees; 4 species shown. Despite considerable variation in size and leaf shape, all oaks are readily distinguished by the presence of *acorns* — oval or roundish, thin-shelled nuts that fit into cuplike caps. Oaks are divided into 2 subgroups: (1) red oaks, which have bristle-tipped leaf lobes or teeth, hairy linings on the insides of the nutshells, and bitter yellow nutmeats; and (2) white oaks, which have leaves that lack bristle tips, smooth linings on the insides of the nutshells, and relatively sweet white nutmeats. The mild acorns of the white oaks are preferable to the more bitter ones of the red oaks, but acorns from both groups can be used.

Use: Nuts, flour or meal, candy. Although a few white oaks have acorns sweet enough to be eaten raw or roasted, most oaks have extremely bitter acorns. Happily, the bitterness is due to an abundance of tannin which is readily soluble in water. Whole kernels, stripped of their shells and boiled in repeated changes of water until the water no longer turns brown, can be roasted and eaten as nuts or dipped in sugar syrup and eaten as candy. Dried and crushed acorns can be placed in porous bags and put through same boiling process to remove the tannin. They can then be redried, ground into meal, and used to make excellent breads and muffins. Rich in protein and fat.

Early Fall

BLACK OAK, *Q. velutina.* A member of the red oak group. Leaves 4–10 in. (10–25 cm) long; moderately lobed; somewhat hairy beneath, *glossy* above. Buds *sharply angled* and *gray-hairy.* Height 70–80 ft. (21–24 m); diameter 3–4 ft. (0.9–1.2 m). Bark is dark, blocky. **Where found:** Dry soils. S. Minn. to s. Maine, south to e. Texas, nw. Fla.

LIVE OAK, *Q. virginiana.* A spreading, southern tree or shrub often hung with Spanish Moss. A member of the white oak group. Leaves thick, leathery, *evergreen,* usually with *rolled edges.* Height to 60 ft. (18 m); diameter to 8 ft. (2.4 m). **Where found:** Mostly coastal plain, but inland toward Southwest. Se. Va. to s. Fla., west to cen. Texas and sw. Okla.

CHESTNUT OAK, *Q. prinus.* Leaves 4–9 in. (10–22.5 cm) long, with 7–16 pairs of *rounded teeth.* Trunk bark distinctive, dark, deeply ridged. Height 60–70 ft. (18–21 m); diameter 3–4 ft. (0.9–1.2 m). **Where found:** Dry woods. S. Ill., s. Ontario, N.Y., sw. Maine, south to ne. Miss., n. Ala., nw. Ga., se. Va.

WHITE OAK, *Q. alba.* **Color pl. 13.** Bark light-colored, slightly furrowed to scaly. Leaves 2–9 in. (5–22.5 cm) long; evenly lobed. Acorn cup is bowl-shaped, covers less than $\frac{1}{3}$ of acorn. Height 60–80 ft. (18–24 m); diameter 2–3 ft. (60–90 cm). **Where found:** Dry or moist soil. Minn. to Maine, south to e. Texas, nw. Fla.

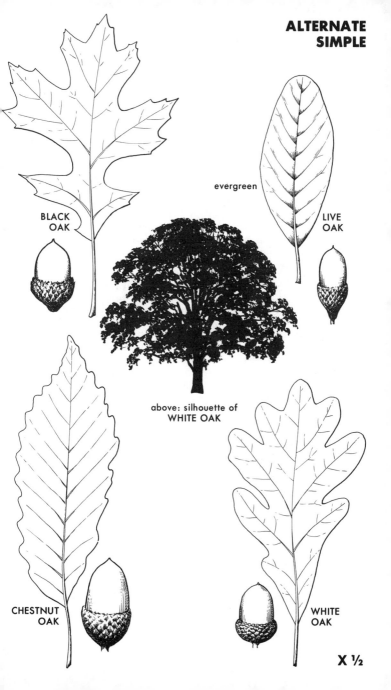

ALTERNATE SIMPLE

BLACK OAK

evergreen

LIVE OAK

above: silhouette of WHITE OAK

CHESTNUT OAK

WHITE OAK

X ½

SHRUBS WITH AROMATIC, TOOTHED LEAVES

BAYBERRIES, WAX-MYRTLE   **Leaves and nutlets**
Myrica spp.
Low, dense shrubs or small trees with stiff branches bearing oblong or narrower, toothless or slightly toothed leaves which are 1–4 in. (2.5–10 cm) long. Note the clusters of small, round, berrylike nutlets covered with white or gray *wax*. Leaves and nutlets *strongly aromatic.* Several species in our area; 2 shown.
Use: Seasoning, candles. The leaves and nutlets can be used in place of commercial bay leaves. The wax boiled from the nutlets can be made into aromatic candles.
                    SUMMER (leaves); LATE SUMMER–FALL (nutlets)
NORTHERN BAYBERRY, *M. pensylvanica.* A northern shrub with *non-evergreen* leaves that are oblong and pointed at the tips. Resin-dots few or missing on upper surfaces of leaves. Twigs *gray-hairy.* **Where found:** Poor soils, mostly near coast and Great Lakes. N. Ohio, s. Newfoundland, and e. New Brunswick, south to N.C. **Fruit:** June–next April.
COMMON WAX-MYRTLE, SOUTHERN BAYBERRY, *M. cerifera.* An *evergreen,* southern shrub or small tree. Leaves often leathery; *wedge-shaped;* toothed or not toward tip; resin-dots both above and below. Twigs nearly hairless. Height 10–30 ft. (3–9 m). **Where found:** Coastal plain, moist sandy or gravelly soil. S. N.J. to Fla., west to e. Texas, and north to Ark. **Fruit:** Aug.-Oct.

SWEETGALE   **Leaves, nutlets**
Myrica gale   **Color pl. 5**
A low northern shrub with *aromatic,* deciduous leaves. Leaves thin, wedge-shaped, blunt-tipped, grayish green with prominent resin-dots beneath; toothed at tip, 1–3 in. (2.5–7.5 cm) long. Fruit compact little cones of tiny, aromatic nutlets; not wax-covered. Height to 6 ft. (1.8 m). **Where found:** Swamps, shallow water. Canada, n. U.S.; in mts. to Tenn. and N.C. **Fruit:** July–winter.
Use: Tea, seasoning. The dried leaves make a delicate tea when steeped in hot water. Both the leaves and aromatic nutlets can be used as a sagelike seasoning for meats.   SUMMER

SWEETFERN   **Leaves**
Comptonia peregrina
A low deciduous bush with strongly *aromatic, fernlike* leaves. Leaves 3–6 in. (7.5–15 cm) long; fruit small, non-waxy nutlets. Height to 5 ft. (1.5 m). **Where found:** Dry soil. Manitoba to Nova Scotia, south to Minn., ne. Ill., nw. Ind., Ohio, W. Va., and Va.; in mts. to n. Ga. **Fruit:** Sept.-Oct.
Use: Tea. The spicy dried leaves can be brewed into a pleasant tea.   SUMMER

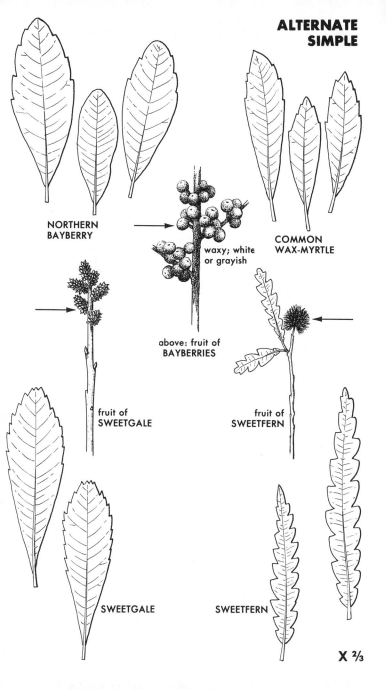

ALTERNATE SIMPLE

NORTHERN BAYBERRY

COMMON WAX-MYRTLE

waxy; white or grayish

above: fruit of BAYBERRIES

fruit of SWEETGALE

fruit of SWEETFERN

SWEETGALE

SWEETFERN

X ⅔

SHRUBS WITH AROMATIC,
TOOTHLESS LEAVES

Note: See also Sassafras, p. 210, Creeping Snowberry, p. 222.

COMMON SPICEBUSH   **Young leaves, twigs, bark;**
Lindera benzoin   **Color pl. 5**   **fruit**
A slender, spicy-scented shrub. Leaves 2–6 in. (5–15 cm) long,
aromatic, thin, elliptic, toothless, and hairless or nearly so.
Bark smooth; twigs slim and brittle. Clusters of spicy-scented
yellowish flowers precede leaves in spring. Fruit an oval, aro-
matic, reddish berry with a single oval seed. Height up to 12 ft.
(3.6 m). **Where found:** Damp woods, streamsides. Se. Kans.,
Iowa, s. Mich., s. Ontario, and sw. Maine, south to Texas and
Fla. **Flowers:** March–May. **Fruit:** July–Sept.
Use: Tea, seasoning. A pleasantly aromatic tea can be made by
steeping the young *leaves, twigs, and bark* for about 15 min.
Dried and powdered, the oily *berries* can be used as a substitute
for allspice.   ALL YEAR (twigs, bark);
                                          SPRING–SUMMER (leaves);
                              LATE SUMMER–EARLY FALL (fruit)

LABRADOR-TEA   **Leaves**
Ledum groenlandicum   **Color pl. 4**
A low-growing northern shrub. Leaves leathery, evergreen,
with *rolled margins* and white or rusty *wool beneath;* fragrant.
1–2 in. (2.5–5 cm) long. Flowers small, 5-petaled, white; in
showy terminal clusters. Up to 3 ft. (90 cm). **Where found:**
Peaty soils, bogs, alpine areas. Canada, n. U.S. **Flowers:**
May–June.
Use: Tea. The dried leaves make a mild and agreeable tea
when steeped for 5–10 min.   ALL YEAR

REDBAY   **Leaves**
Persea borbonia
An evergreen tree or large shrub. Leaves toothless, narrowly
elliptic, 3–6 in. (7.5–15 cm) long; leathery, glossy bright green
above, pale underneath; *aromatic, smells of bay leaves.* Twigs
angled; bark is dark reddish, deeply grooved. Clusters of small
creamy-white flowers in leaf axils. Fruit blue or black, dusted
with grayish powder, 1-seeded; in red-stemmed clusters. Height
to 60 ft. (18 m); diameter to 3 ft. (90 cm). **Where found:**
Moist, sandy woods, swamps. Coastal plain; s. Del. to Fla., west
to Texas. **Flowers:** May–July. **Fruit:** Aug.–Sept.
Use: Seasoning. Use the fresh or dried leaves in place of commercial bay leaves.   ALL YEAR

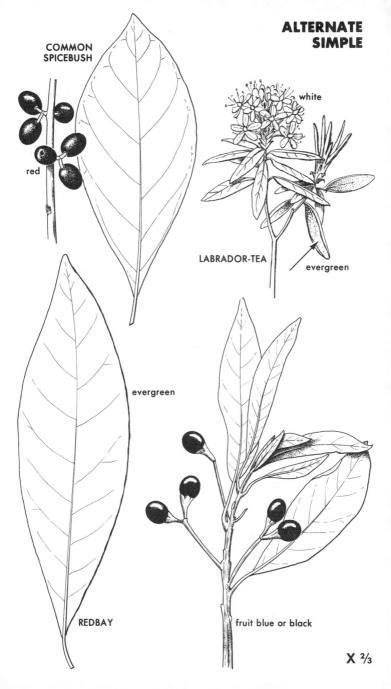

ALTERNATE SIMPLE

COMMON SPICEBUSH

red

white

LABRADOR-TEA

evergreen

evergreen

REDBAY

fruit blue or black

X ⅔

TREES WITH LEAVES OFTEN 2- OR 3-LOBED

SASSAFRAS   **Roots, young leaves**
Sassafras albidum

A medium-sized tree. Leaves 3–9 in. (7.5–22.5 cm) long, tooth-less, ovate but often also 2- or 3-lobed; all 3 leaf shapes usually occur on same tree. Twigs *green,* often branched; mature bark red-brown, furrowed. Crushed leaves, twigs, and bark aromatic, spicy. Fruit small, blue, fleshy, 1-seeded; on red stalks. Height 10–50 ft. (3–15 m); diameter to 1 ft. (30 cm). **Where found:** Old fields, borders of woods. Se. Iowa and cen. Mich. to sw. Maine, south to e. Texas, cen. Fla. **Fruit:** Aug.–Oct.

Use: Tea, seasoning, soup thickener. To make tea, boil the thoroughly washed *roots* until the water turns a rich reddish-brown and sweeten. Excellent. The *young leaves* can be dried, rubbed into a fine powder, and used to make "gumbo filet" or to thicken soup. **Warning:** Sassafras has recently been proven to contain a chemical that causes cancer in laboratory test animals.   ALL YEAR (roots); SPRING–SUMMER (leaves)

RED MULBERRY   **Young shoots,**
Morus rubra   **ripe fruit (only)**

Leaves fine-toothed, 3–6 in. (7.5–15 cm) long, somewhat *sand-papery* above and *hairy* below; often with 2 or 3 lobes. Twigs *hairless.* Sap of twigs and leafstalks *milky.* Trunk bark red-brown with smooth ridges. Fruit pendent, blackberrylike; red at first, purple when ripe. Height to 70 ft. (21 m); diameter to 3 ft. (90 cm). **Where found:** Rich soil, open woods, fencerows. Se. Neb., se. Minn., s. Ontario, N.Y., sw. Vt., and Mass., south to cen. Texas and s. Fla. **Fruit:** May–July.

Use: Fresh fruit, jelly, cold drink, cooked vegetable. A much neglected but delicious *fruit;* pectin is needed when making jelly. The tender new leading *shoots* make an outstanding cooked vegetable when collected as the leaves are just un-folding and boiled for 20 min. **Warning:** Unripe fruit and raw shoots contain hallucinogens.   SPRING (shoots);
                                              EARLY SUMMER (fruit)

WHITE MULBERRY (not shown)   **Young shoots,**
Morus alba   **ripe fruit (only)**

Similar to Red Mulberry, but with *hairless* leaves and *yellow-brown bark.* Fruit *white,* insipid, sweet. **Where found:** Rich soil. Ontario to Maine and south. **Fruit:** May–July.

Use: Dried fruit, cooked vegetable. Although often too sweet to be enjoyed fresh, the ripe *berries* are excellent dried and added to nut breads or muffins. Dry the fruit using one of the methods described under Fruit, p. 297. Be sure there is plenty of air circulation; the berries spoil quickly in the presence of moisture. The *young shoots* are prepared like those of Red Mulberry. **Warning:** See Red Mulberry, above.

                SPRING (shoots); EARLY SUMMER (fruit)

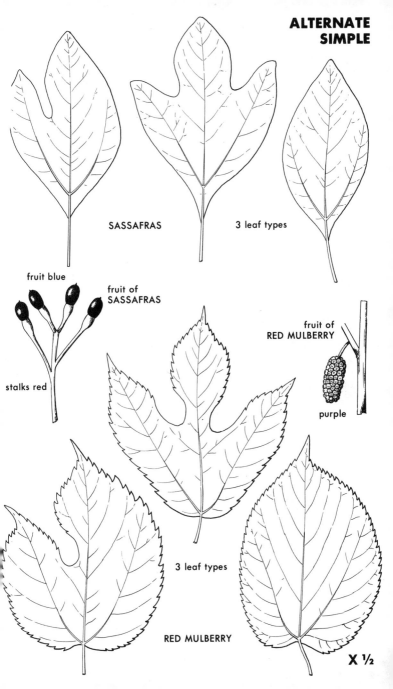

ALTERNATE SIMPLE

SASSAFRAS

3 leaf types

fruit blue

fruit of SASSAFRAS

stalks red

fruit of RED MULBERRY

purple

3 leaf types

RED MULBERRY

X ½

GOOSEBERRIES and CURRANTS   Fruit
Ribes spp.

Low, erect, or sprawling shrubs with 3- to 5-lobed maplelike leaves that are generally ½–5 in. (1.3–12.5 cm) long; leaf edges coarsely toothed. Leaves may be clustered on short "spur" branches. Small greenish-white to purple, 5-petaled flowers cluster on long stalks originating at bases of leafstalks. Flowers followed by nearly round, juicy, many-seeded fruit that are yellow, red, or purple to black. *Ribes* species are divided into 2 subgroups: (1) gooseberries, which have 1–3 *thorns* at bases of leafstalks and clusters of 1–5 *bristly* fruit firmly joined to their stalks; and (2) currants, which usually *lack thorns* and have lengthened clusters of numerous *smooth-skinned* fruit that separate easily from their stalks. Numerous species in our area. GARDEN RED CURRANT, *R. sativum,* and PASTURE GOOSEBERRY, *R. cynosbati,* shown. Height up to 5 ft. (1.5 m). **Where found:** Openings in woods, fields; some species in rocky areas, others prefer swamps. One or more species can be found through most of our area. **Flowers:** April–Aug. **Fruit:** June–Sept.

Use: Fresh or dried fruit, jelly, pie filling, fruit sauce. Although all gooseberries and currants produce edible fruit, the quality varies from species to species. Many are sweet and delicious eaten fresh; others are extremely tart and must be cooked with sugar. Normally, gooseberries should be cooked to reduce their spines. All species are rich in pectin.   SUMMER

PURPLE-FLOWERING RASPBERRY   Fruit
Rubus odoratus   Color pl. 12

A thornless shrub with maplelike leaves and showy rose-purple, roselike flowers that are 1–2 in. (2.5–5 cm) in diameter. Note the reddish brown, sticky-hairy stems. The shallow, cup-shaped red berries are dry and sour. Height 3–6 ft. (0.9–1.8 m). THIMBLEBERRY, *R. parviflorus* (not shown), is similar but has white flowers and rarely sticky stems; w. Ontario, n. Mich., Minn., and west. **Where found:** Rocky woods, thickets, ravines. S. Ontario, s. Quebec, Nova Scotia, south to Tenn. and Ga. **Flowers:** June–Sept. **Fruit:** July–Sept.

Use: Fruit. The berries are rather tart and dry, but quite edible. See other raspberries, p. 184.   SUMMER

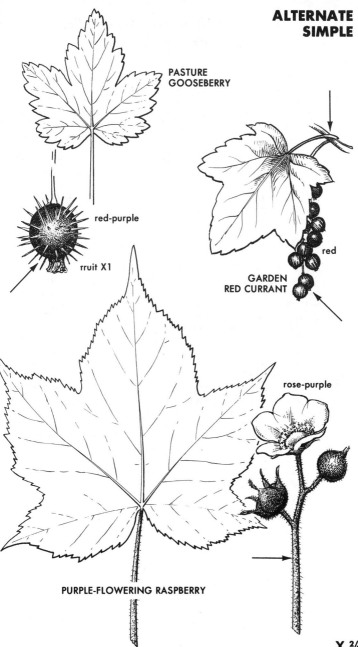

ALTERNATE SIMPLE

PASTURE GOOSEBERRY

red-purple

rruit X1

GARDEN RED CURRANT

red

rose-purple

PURPLE-FLOWERING RASPBERRY

X ⅔

TREES WITH MAPLELIKE LEAVES;
FRUIT DRY PENDENT BALLS

SWEETGUM   **Hardened sap**
Liquidambar styraciflua
A tall tree with distinctive star-shaped, hairless, *5- to 7-lobed* leaves. Leaves pleasantly aromatic when crushed; 5–8 in. (12.5–20 cm) long. Mature bark grayish, regularly grooved. Twigs often with corky wings. Fruit a dry, brown, somewhat prickly ball hanging by a long stem. Height to 120 ft. (36 m); diameter to 5 ft. (1.5 m). **Where found:** Rich, wet soil; lowland woods. Se. Mo., s. Ill., s. Ohio, W. Va., se. N.Y., and s. Conn., south to e. Texas and cen. Fla. **Fruit:** Sept.–Nov. (or later).
Use: Chewing gum. The aromatic, hardened sap that exudes from wounds in the tree has long been used as a substitute for chewing gum in the South.   ALL YEAR

SYCAMORE   **Sap**
Platanus occidentalis
A tall, massive tree with a wide-spreading crown. Bark brittle, flaking off in large irregular patches revealing whitish underbark. Leaves *3- to 5-lobed,* coarsely toothed, almost hairless, 6–10 in. (15–25 cm) long. Flowers small, in rounded heads. Fruit tiny, hairy; in long-stalked compact balls ¾–1½ in. (1.9–3.8 cm) wide; one fruit ball to a stalk. Height to 130 ft. (39 m); diameter to 8 ft. (2.4 m). **Where found:** Streambanks, bottomlands. E. Neb., Iowa, cen. Mich., s. Ontario, N.Y., and sw. Maine, south to cen. Texas and nw. Fla. **Flowers:** April–May. **Fruit:** Oct.–spring.
Use: Syrup, sugar, water. Sycamores can be tapped and the sap boiled down into syrup and sugar, just as with the maples (p. 176). Unfortunately, huge amounts of sap are needed, and the results are mediocre. However, the sap is a fine source of pure drinking and cooking water in areas with contaminated water.   LATE WINTER–EARLY SPRING

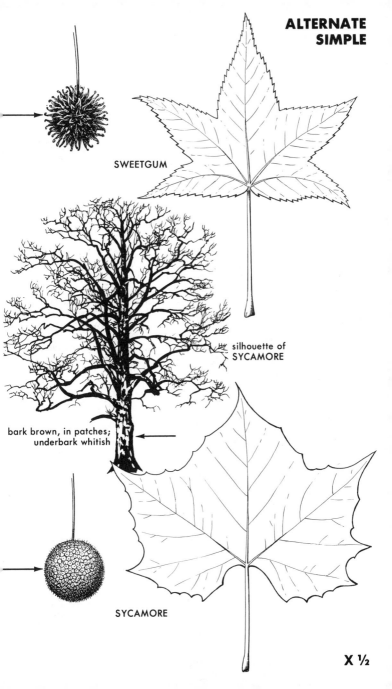

ALTERNATE SIMPLE

SWEETGUM

silhouette of SYCAMORE

bark brown, in patches; underbark whitish

SYCAMORE

X ½

THORNY OR SPURRED SMALL TREES; APPLELIKE FRUIT

CRABAPPLES   Fruit
Pyrus spp.   **Color pl. 12 (Siberian Crab)**
Similar to Domestic Apple (below) but leaves sharply toothed; leaf undersides of some species hairless. Fruit generally smaller than cultivated apples, but often as much as 1–2 in. (2.5–5 cm) across; yellow to reddish. A variety of species; AMERICAN CRABAPPLE, *P. coronaria,* shown. **Where found:** Old fields, thickets. Much of our area. **Flowers:** March–May. **Fruit:** Sept.–Nov.
Use: Jelly, preserves. Although usually too hard and tart to be enjoyed raw, crabapples contain an abundance of pectin and make excellent jellies or preserves.   FALL

DOMESTIC APPLE (not shown)   Fruit
Pyrus malus
A small tree with familiar fruit. Leaves 1–4 in. (2.5–10 cm) long, ovate, somewhat scalloped-toothed, often *gray- or white-woolly beneath.* Leaves and fruit on short, slightly woolly, spurlike twigs that may have thornlike tips. Bark brownish, scaly. Showy, fragrant, pinkish-white, 5-petaled flowers produce fruit at least 1 in. (2.5 cm) across. 20–30 ft. (6–9 m). **Where found:** Escaped from cultivation. Most of our area. **Flowers:** April–June. **Fruit:** Sept.–Nov.
Use: Jelly, fruit, cold drink. Usually a little more tart than orchard apples. An excellent source of pectin.   FALL

DOMESTIC PEAR (not shown)   Fruit
Pyrus communis
Similar to Domestic Apple (above) but mostly hairless. Leaves 1–3 in. (2.5–7.5 cm) long. Flowers white. Fruit familiar, yellowish, fleshy. 20–35 ft. (6–10.5 m). **Where found:** Spread from cultivation, thickets. S. Canada, ne. U.S.
Use: Fresh or cooked fruit.   FALL

HAWTHORNS   Fruit
Crataegus spp.
Dense shrubs or small trees with *long thorns* — 1–5 in. (2.5–12.5 cm) — and smooth or scaly bark. Branches crooked, irregular. Leaves lobed or not, sharply toothed. Flowers 5-petaled, white or pale pink; fruit usually red, somewhat resembling rose hips or tiny apples. Numerous species; most are difficult to differentiate. Typical form shown. Up to 25 ft. (7.5 m). **Where found:** Wet, dry, or rocky woods; old fields. One or more species found in most of our area. **Flowers:** April–June. **Fruit:** Oct.
Use: Jam, jelly, tea. Although not usually good raw, the small fruit make an excellent jelly; they contain plenty of their own pectin. To make a pleasant tea, steep the fruit along with a little Peppermint (pp. 118, 138).   FALL

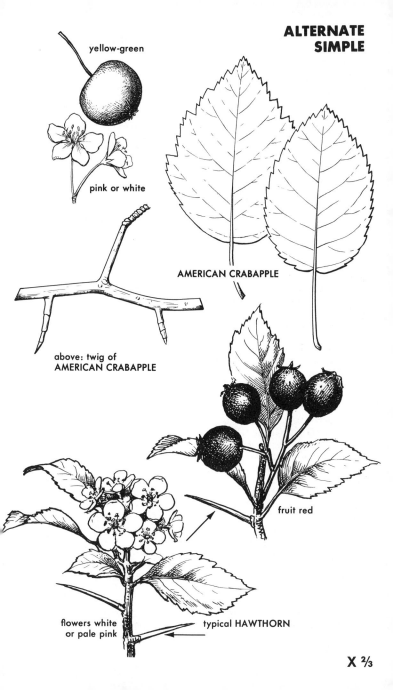

yellow-green

pink or white

AMERICAN CRABAPPLE

above: twig of
AMERICAN CRABAPPLE

fruit red

flowers white
or pale pink

typical HAWTHORN

X ⅔

CHERRIES AND PLUMS

WILD PLUMS   **Fruit**
Prunus spp.

Shrubs or small trees similar to wild cherries (below) but occasionally thorny. Fruit larger, powdered with white, encircled by a line along the long axis; large solitary seeds usually *somewhat flattened.* Fruit yellow, red, purplish, or blackish. A variety of species found in thickets throughout our area.

Use: Fruit, jelly. Fruit quality varies from plant to plant; some good fresh, others better cooked. Pectin is needed when making jelly.   LATE SUMMER–EARLY FALL

BEACH PLUM, *P. maritima.* A low, straggling shrub. Leaves 2–3 in. (5–7.5 cm) long, ovate. Twigs, buds, and undersides of leaves downy. Older branches and tops of leaves hairless. Flowers white, appearing before leaves. Fruit purplish. Up to 8 ft. (2.4 m). **Where found:** Sandy coastal areas. Maine to Del. **Flowers:** April–June. **Fruit:** Sept.–Oct.

WILD CHERRIES   **Fruit**
Prunus spp.

Thornless trees or shrubs with clusters of 5-petaled white flowers followed by small, *glossy,* red to black fruit with nearly *round* solitary seeds. Leaves usually narrow-based, singly-toothed, often with long-pointed tips. Bark marked with numerous *horizontal streaks.* Various species; 3 shown.

Use: Fresh fruit, jelly, pies. A few wild cherries can be eaten fresh, but most are best when made into jelly; pectin must be added. Once pitted, the sweeter fruit can be used in pies, pancakes, or muffins. **Warning:** The wilted leaves and fresh seeds contain cyanide and should not be eaten. However, as cooking destroys the cyanide, pitting is unnecessary during the early stages of making jelly.   LATE SUMMER–EARLY FALL

CHOKE CHERRY, *P. virginiana.* Smaller than following species. Leaves ovate, sharp-toothed; midribs hairless. Bark gray-brown, smoothish. Ripe fruit purplish, astringent. To 30 ft. (9 m). **Where found:** Thickets. Canada south to Kans., Md.; in mts. to Ga. **Flowers:** April–July. **Fruit:** July–Oct.

BLACK CHERRY, *P. serotina.* Small to large. Leaves 2–6 in. (5–15 cm) long, narrow, finely blunt-toothed; midrib hairy beneath. Young bark smooth, dark reddish-brown to black; old bark scaly. Flowers and fruit in long clusters; ripe fruit blackish. Up to 80 ft. (24 m). **Where found:** Woods, thickets. S. Canada south. **Flowers:** April–June. **Fruit:** July–Oct.

FIRE or PIN CHERRY, *P. pensylvanica.* A shrub or small tree. Leaves narrow; flower clusters short, umbrellalike. Bark red-brown, smoothish. Ripe fruit red, $\frac{1}{4}$ in. (6 mm) across. Up to 40 ft. (12 m). **Where found:** Burned areas, young woods, thickets. Canada, n. U.S.; in mts. to e. Tenn. and n. Ga. **Flowers:** March–July. **Fruit:** July–Sept.

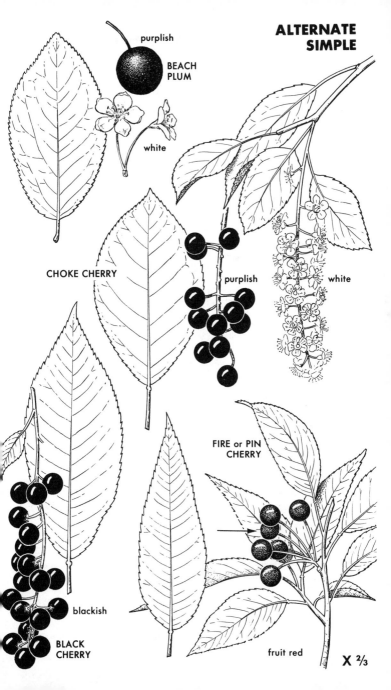

ALTERNATE SIMPLE

purplish

BEACH PLUM

white

CHOKE CHERRY

purplish

white

FIRE or PIN CHERRY

blackish

BLACK CHERRY

fruit red

X ⅔

BLUEBERRYLIKE FRUIT

BLUEBERRIES, HUCKLEBERRIES,   **Fruit**
BILBERRIES, and DEERBERRIES
Vaccinium spp. and *Gaylussacia* spp.
Color pl. 4 (Highbush Blueberry)
Common low to tall shrubs. Leaves elliptic, short-stalked,
toothless or minutely toothed. Twigs slender, greenish or reddish, often zigzag. Flowers bell-like; whitish, pinkish, or greenish. Berries with 5 calyx lobes forming a star pattern; blue or
black, glossy or powdered with white. Typical blueberry
shown. **Where found:** Wet or dry acid soils; bogs, barrens,
tundras, woods, thickets. One or more species found throughout. **Flowers:** April–June. **Fruit:** June–Sept.
Use: Fresh, cooked, or dried fruit; jelly.   Summer

BUCKTHORNS   **Poisonous**
Rhamnus spp.
Shrubs, small trees. Leaves elliptic or oval, fine-toothed,
strongly veined. Berries clustered along twigs; juicy, black,
bitter; *lack prominent calyx lobes.* Several species; 1 shown.
Warning: Leaves and fruit strongly cathartic.
COMMON BUCKTHORN, *R. cathartica. Thorny.* To 20 ft.
(6 m). **Where found:** Thickets. Canada south to Mo., Va.

CHOKEBERRIES   **Fruit**
Pyrus spp.
Common shrubs. Leaves fine-toothed, elliptic, with *minute
dark glands* along midrib on upper side. Flowers white or
pinkish, 5-petaled. Fruit suggest juneberries (below), but clusters *flat-topped;* black, purple, or red. 3 species; 1 shown.
Use: Fruit, jelly. Excellent; use like blueberries.   Fall
BLACK CHOKEBERRY, *P. melanocarpa.* Leaves and twigs
hairless. Fruit *black.* 2–6 ft. (0.6–1.8 m). **Where found:** Wet
to dry thickets, swamps. Minn., nw. Ontario, Newfoundland,
south to Tenn., S.C. **Flowers:** April–June. **Fruit:** Aug.–Oct.
RED CHOKEBERRY, *P. arbutifolia* **Color pl. 5 (only).**
Twigs, leaf undersides woolly. Fruit *red.* 3–8 ft. (0.9–2.4 m).
Where found: Swamps, bogs. Mo., Mich., Nova Scotia, south
to e. Texas, Fla. **Flowers:** April–July. **Fruit:** Sept.–Nov.

JUNEBERRIES, SERVICEBERRIES   **Fruit**
Amelanchier spp.
Shrubs or small trees. Leaves oval, sharp- or blunt-tipped,
toothed. Bark tight, grayish. Flowers white, 5-petaled, in
drooping clusters; often precede leaves. Fruit purple-black.
Numerous species; many with sweet, juicy fruit. Typical form
shown. **Where found:** Thickets, woods. Canada south to Ga.,
Ala. **Flowers:** April–June. **Fruit:** June–Sept.
Use: Fruit, jelly. An excellent and far too neglected fruit; use
like blueberries.   Summer (varies by species and location)

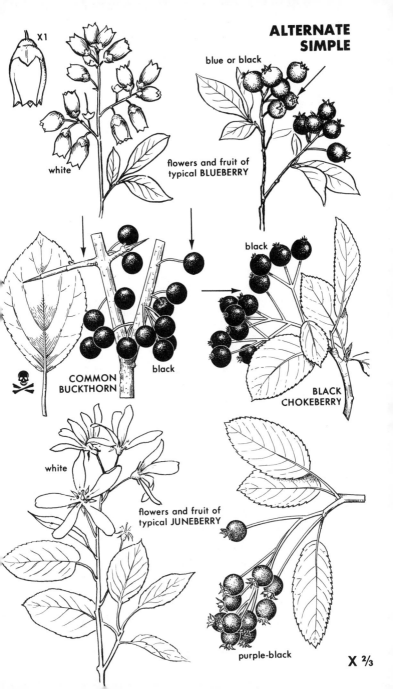

ALTERNATE SIMPLE

X1

white

blue or black

flowers and fruit of typical BLUEBERRY

COMMON BUCKTHORN

black

black

BLACK CHOKEBERRY

white

flowers and fruit of typical JUNEBERRY

purple-black

X ⅔

TINY, OVAL, EVERGREEN LEAVES; CREEPING OR MAT-FORMING SHRUBS

CRANBERRIES   Fruit
Vaccinium spp.

Low creeping northern shrubs with small, alternate, oval, ever-green leaves. Leaves hairless, odorless. Flowers small, pink. Fruit many-seeded juicy red berries. 3 species shown.
Use: Cooked fruit, jelly, cold drink. Substitute wild cranber-ries in any recipe calling for commercial cranberries; the berries contain their own pectin. The berries are usually gathered while still firm, just before, or just after, the first frost in the fall, but they often remain on the stems during the winter. Cranberries keep for some time. The thoroughly washed *firm* berries can be packed in *sterilized* jars and kept in a refrigerator for months. The berries of Mountain Cranberry are mildly bitter but just as good as regular cranberries when cooked.

FALL–WINTER

MOUNTAIN CRANBERRY, *V. vitis-idaea.* Leaves blunt, *dotted with black beneath.* Flowers pink, bell-like. Up to 7 in. (17.5 cm). **Where found:** Exposed rocky or dry peaty soil. Canada south to n. Minn. and mts. of n. New England. **Flow-ers:** June–July. **Fruit:** Aug.–Oct. (holding over winter).

SMALL CRANBERRY, *V. oxycoccus.* Smaller than the fol-lowing species, with *pointed* leaves that are white beneath and have rolled edges. Fruiting stalks spring from tips of stems. Berries slightly smaller. **Where found:** Bogs or wet, peaty soil. Canada south to Minn., Mich., n. Ohio, N.J.; in mts. to N.C. **Flowers:** May–July. **Fruit:** Aug.–Oct. (holding over winter).

LARGE CRANBERRY, *V. macrocarpon.* **Color pl. 4.** *Blunt* leaves. Fruiting stalks not from tip of stem. **Where found:** Open bogs. Minn. to Newfoundland and south to Ill., Ohio, Va. **Flowers:** June–Aug. **Fruit:** Sept.–Nov. (holding over winter).

CREEPING SNOWBERRY, MOXIE-PLUM   Leaves,
Gaultheria hispidula   **Color pl. 15.**   fruit

A ground-hugging, mat-forming, northern plant with very small — $\frac{1}{4}$-$\frac{1}{2}$ in. (0.6–1.3 cm) — alternate, oval leaves along creeping stems. *Leaves smell of wintergreen* when crushed. Small, white flowers and berries are in leaf axils. **Where found:** Mossy (usually evergreen) woods, bogs, logs. Canada, n. U.S., south locally in Appalachians. **Fruit:** Aug.–Sept.
Use: Tea, fresh or cooked fruit. The fresh *leaves* make an excellent tea. The juicy, wintergreen-flavored *berries* (often hidden beneath the leaves) are excellent eaten fresh with cream and sugar. They can also be made into fine jams and preserves with the addition of pectin.   ALL YEAR (leaves);

LATE SUMMER (fruit)

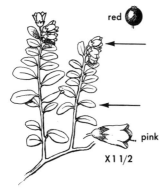

MOUNTAIN CRANBERRY

SMALL CRANBERRY

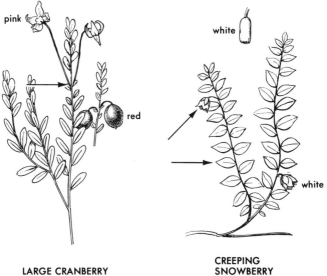

LARGE CRANBERRY

**CREEPING
SNOWBERRY**

X ⅔

LOW, CREEPING OR TRAILING PLANTS; LEAVES EVERGREEN, LEATHERY

WINTERGREEN, CHECKERBERRY — Leaves, fruit
Gaultheria procumbens

A low evergreen plant that spreads by slender underground runners. Leaves thick, shiny, oval, slightly toothed; 1–2 in. (2.5–5 cm) long. Crushed leaves *smell of wintergreen*. Small, waxy, egg-shaped flowers dangle beneath the leaves. Fruit a small wintergreen-flavored red berry. 2–5 in. (5–12.5 cm). **Where found:** Poor soil, woods, clearings. Canada, n. U.S., south in mts. to Ala. and Ga. **Flowers:** July–Aug. **Fruit:** Aug.–next June.

Use: Tea, fruit, salad. The *leaves* can be gathered throughout the year and used to make an excellent tea. Both the tender new leaves and *berries* can be used as trailside nibbles or added to salads. The berries often remain on the plant until the next flowering season. ALL YEAR (leaves); FALL–EARLY SPRING (fruit)

BEARBERRY, KINNIKINIK — Fruit, leaves
Arctostaphylos uva-ursi

A low trailing shrub with papery reddish bark and small *paddle-shaped,* evergreen leaves. Flowers pink or white, in terminal clusters; egg-shaped, with small, lobed mouths. Fruit a red berry. **Where found:** Exposed rock or sand. Arctic region of Canada, south locally to n. U.S. **Flowers:** May–July.

Use: Cooked fruit; tobacco. Although dry and not particularly inviting raw, the *berries* are quite pleasant when cooked and served with cream and sugar. The dried *leaves* are a tobacco substitute. SUMMER (leaves); LATE SUMMER–FALL (fruit)

TRAILING ARBUTUS — Corolla (flower tube)
Epigaea repens

A low trailing shrub with *oval, leathery,* evergreen leaves that are 1–5 in. (2.5–12.5 cm) long. Stem normally brown-hairy. Flowers pink or white, clustered, tubular, with 5 flaring lobes. **Where found:** Dry soil; woods, clearings. Canada, n. U.S., south in mts. to Ga. **Flowers:** Late Feb.–May.

Use: Nibble, salad. The raw corolla, or flower tube, makes an excellent sour-sweet nibble or addition to salads. **Note:** Protected in many states. EARLY SPRING

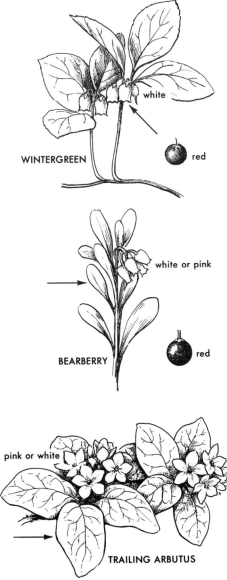

WINTERGREEN

white

red

BEARBERRY

white or pink

red

pink or white

TRAILING ARBUTUS

X ⅔

Miscellaneous Plants

LARGE, COARSE GRASSES; DENSE STANDS

LARGE CANE   **Young shoots, seeds**
Arundinaria gigantea
Extremely tall, woody, hollow-stemmed, *bamboolike;* often
forms extensive canebrakes in South. Upper stems of older
plants branched. Leaves flat, grasslike; up to 18 in. (45 cm)
long, 1½ in. (3.8 cm) wide. Loose-branched clusters of flattened
seed-bearing spikelets produced at long irregular intervals from
leafy branches of older stems. 5–30 ft. (1.5–9 m). **Where
found:** Low ground, riverbanks, swamps. Mo. to Del. south.
Use: Asparagus, cereal, flour. The *young shoots* can be prepared like bamboo shoots. The large *seeds* can be used as a
cereal or ground into flour. **Warning:** See Wild Rice, below.
EARLY SPRING (shoots); SUMMER (seeds)

WILD RICE   **Seeds**
Zizania aquatica   **Color pl. 3**
A tall, graceful, aquatic grass. Leaves light green, lance-shaped;
1–3 ft. (30–90 cm) long, 1–2 in. (2.5–5 cm) wide. Flower cluster
to 2 ft. (60 cm) long; pistillate flowers on erect broomlike
branches at summit, staminate flowers dangle from spreading
side branches below. Seeds slender, cylindrical; encased in
papery, bristle-tipped husks. Up to 10 ft. (3 m). **Where found:**
Shallow water. Minn. to New Brunswick south.
Use: Cereal, flour. Using a boat, collect the ripening grain just
before it drops of its own accord. Bend the stalks over a sheet
and rap them with a stick. Dry thoroughly (parch) after har-
vesting, rub gently to break up husks, and winnow. Wash in
cold water to remove smoky flavor. Then prepare like brown
rice or grind into flour. **Warning:** Although infrequent,
ERGOT, *Claviceps* spp., highly poisonous *pink or purplish*
fungi, can replace some of the seeds. When hardened, they
assume the seed's shape and may be the same size, or 3–4 times
larger. If present, collect in another area.
MID-LATE SUMMER

REED, PHRAGMITES   **Young stems,**
Phragmites communis   **seeds, rootstock**
Note the silky, plumelike, terminal flower cluster; purplish
when young, whitish and fluffy when older. Stems conspicuously jointed, hollow or pithy, leafy. Leaves gray-green; to 2 ft.
(60 cm) long, 2 in. (5 cm) wide. Rootstock long, creeping. 6–10
ft. (1.8–3 m). **Where found:** Fresh or brackish marshes,
ditches. Canada south to Texas, La., Ill., Ohio, Md.
Use: Candy, cereal, flour. *Young stems,* while still green and
fleshy, can be dried and pounded into a fine powder, which
when moistened is roasted like marshmallows. The tiny reddish *seeds* can be ground into flour or made into gruel. The
rootstocks can be crushed and washed to obtain flour (see
cattails, p. 158).   EARLY SUMMER (stems); FALL (seeds);
FALL–EARLY SPRING (rootstock)

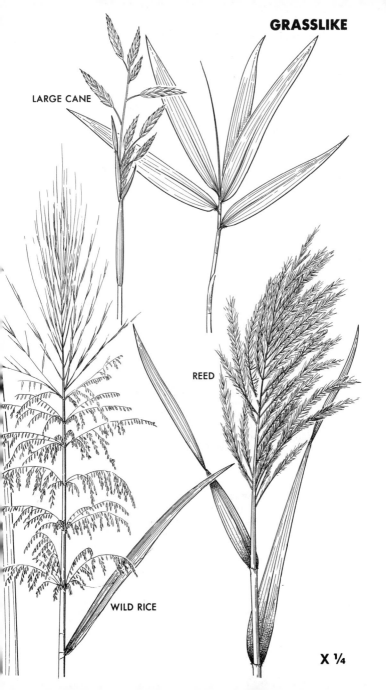

GRASSLIKE

LARGE CANE

REED

WILD RICE

X ¼

MISCELLANEOUS GRASSLIKE PLANTS

GREAT BULRUSH, TULE   Shoots, pollen,
Scirpus validus and *S. acutus*   seeds, rootstock
Color pl. 3

Tall, dark green marsh plants in dense stands. Stems smooth, round, pithy, leafless; topped by branching clusters of brown, bristly flower spikes. Rootstock thick, scaly. *S. validus* shown. 3–9 ft. (0.9–2.7 m). **Where found:** Mud, shallow fresh or brackish water. **Throughout. Flowers:** May–Sept.

Use: Salad, cooked vegetable, potato, flour. The *young shoots* are excellent raw or cooked, as are the tender cores at the bases of older shoots. The *pollen* and ground-up *seeds* can be used as flour. The leading tips of the *rootstocks* are rich in starch and sugar and can be roasted several hours and eaten like potatoes, or dried and pounded into flour. SPRING (shoots); SUMMER (pollen); FALL (seeds); FALL–EARLY SPRING (rootstock)

CHUFA, YELLOW NUT-GRASS   Tubers
Cyperus esculentus

Note the feathery radiating flower cluster bearing numerous yellowish spikelets. Stem 3-sided. Leaves light green, grasslike; basal and in a whorl at base of flower cluster. Slender, scaly runners terminated by small nutlike tubers radiate from base of plant. 1–3 ft. (30–90 cm). **Where found:** Damp sandy soil, waste places. S. Canada south. **Flowers:** June–Oct.

Use: Nibble, cooked vegetable, flour, coffee, cold drink. Excellent raw, boiled, dried and ground into flour, or roasted to a dark brown and ground into coffee. To make a cold drink, soak the tubers in water for several days, then crush in fresh water, strain through cheesecloth, and sweeten. ALL YEAR

GREAT BUR-REED   Tubers
Sparganium eurycarpum

Aquatic. Staminate (whitish) and pistillate flowers in *separate spherical heads* along top of stem. Stem stout, erect, simple or sparingly branched, often zigzag. Basal leaves long, stiffish, bladelike; stem leaves alternate. Fruit green, in burlike spheres. Creeping rootstocks produce small tubers in autumn. 2–5 ft. (0.6–1.5 m). **Where found:** Mud, shallow water. S. Canada south to Kans., N.J. **Flowers:** May–Aug.

Use: Potato. Tubers usually widely scattered and difficult to gather in quantity. FALL–EARLY SPRING

CATTAILS   Young shoots and stalks, immature
         flower spikes, pollen, sprouts, rootstock
Typha spp.   **Color pl. 3**

Tall marsh plants with swordlike leaves, stiff stems, and brown sausagelike flowerheads. See p. 158.

Use: Salad, asparagus, cooked vegetable, flour, potato. See p. 158.   ALL YEAR

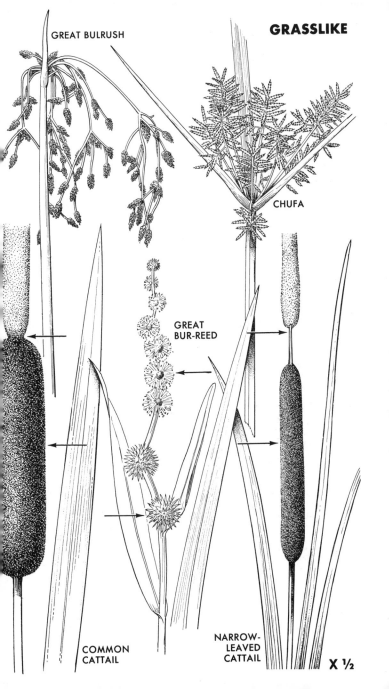

GREAT BULRUSH

GRASSLIKE

CHUFA

GREAT
BUR-REED

COMMON
CATTAIL

NARROW-
LEAVED
CATTAIL

X ½

FERNS

Note: The two species shown are the commonest and tastiest species found in our area. While the fiddleheads of other ferns can presumably be eaten (none are reported to be poisonous), many are bitter and unpalatable.

Caution: Do not mistake the young shoots of Poison Hemlock (p. 38) or Water-hemlock (p. 42) for the fiddleheads of one of the smaller ferns.

BRACKEN, PASTURE BRAKE   **Fiddlehead**
Pteridium aquilinum
Our most abundant fern. Note the large, coarse, erect fronds rising singly from long slender rootstocks; *broadly triangular* in outline, 1–3 ft. (30–90 cm) wide, distinctly 3-forked, with numerous *leathery* oblong leaflets. Lower leaflets cut into blunt-tipped subleaflets. Fiddleheads shaped like an eagle's claw when unfolding, covered with silver-gray hair. Mature stems smooth, rigid, partially grooved in front; green, then brown. 1–6 ft. (0.3–1.8 m). **Where found:** Dry, open, sunny places; woods, old pastures, burns. Throughout.
Use: Salad, asparagus. Gather when 6–8 in. (15–20 cm) high and rub off the woolly covering. Add in small amounts to salads, or boil for 30–45 min. **Warning:** Cooking recommended. Raw plant contains an enzyme which, when ingested in sufficient quantities, destroys vitamin B_1 (thiamine). Mature fronds poisonous to cattle. *Also,* recent evidence from Japan suggests a link between the excessive consumption of Bracken over an extended period of time and a high incidence of stomach cancer.   Spring

OSTRICH FERN   **Fiddleheads**
Pteretis pensylvanica
A large northern fern growing in graceful vaselike clumps from a stout scaly rootstock. Fronds are of 2 types: (1) large, arching, plumelike, dark rich green sterile fronds; and (2) smaller, stiffly erect fertile (fruiting) fronds, resembling thick dark-brown feathers, inside the circle of sterile fronds. Fiddleheads tightly coiled, rich emerald-green; covered with large, papery, brown scales. 2–6 ft. (0.6–1.8 m). **Where found:** Rich, moist or wet soil, some sunshine; streambanks, riverbanks, open woods, edges of swamps. Canada south to Mo., Ohio, n. Va.
Use: Salad, asparagus. Gather when under 6 in. (15 cm) high and still tightly curled. Excellent added to salads or boiled for 10–15 min.   Early Spring

232

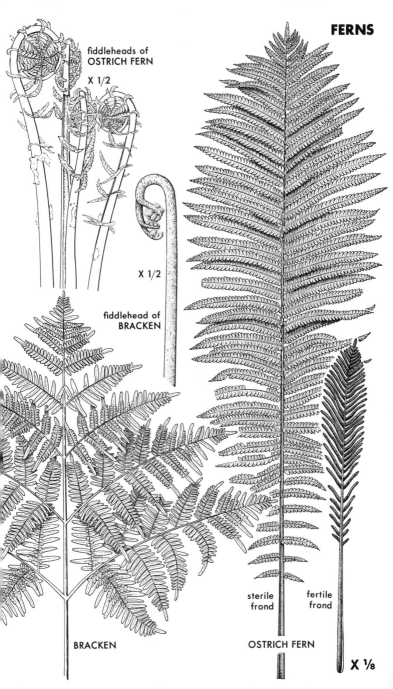

FERNS

fiddleheads of
OSTRICH FERN
X 1/2

X 1/2

fiddlehead of
BRACKEN

BRACKEN

sterile
frond

fertile
frond

OSTRICH FERN

X ⅛

SEAWEEDS

DULSE   **Fronds**
Rhodymenia palmata
Erect, dark purple-red, up to 1 ft. (30 cm) long. Stem short, slender; quickly flaring into a thin, fan- or tongue-shaped, ribless frond. Fronds often cleft into flat fingers which may or may not end in round-tipped lobes. **Where found:** Rocks; low-water mark and below. N.J. northward.
Use: Nibble, seasoning. Becomes pliable and tender when partially dry. Eat as is or grate into seafood casseroles.
ALL YEAR

IRISH MOSS   **Fronds**
Chondrus crispus
Olive-green to purplish (whitish when sun-bleached), in dense beds. Fronds flattened; tough and elastic wet, brittle dry; freely forking; terminal lobes crowded. 3–6 in. (7.5–15 cm). **Where found:** Rocks; low-tide mark. N.C. northward.
Use: Blancmange, soup, jelly, cooked vegetable. Fronds release gelatin when boiled. To make blancmange, thoroughly wash 1 cupful in fresh water, tie in a cheesecloth bag, cook for 30 min. with 2 quarts of milk in a double boiler, sweeten, and chill. Boil in water to make a tasteless but nourishing soup or jelly. Tender when cooked.   ALL YEAR

EDIBLE KELP   **Midrib, lateral fronds**
Alaria esculenta
Stem short, cylindrical. Main frond unperforated, thinnish, wavy-margined, with a strong *midrib;* olive-green or -brown; 1–10 ft. (0.3–3 m) long. Note *tonguelike lateral fronds.* **Where found:** Submerged ledges, washed ashore. Cape Cod north.
Use: Salad, seasoning, soup. Add the thinly sliced *midribs* to salads. Shred the *dried lateral fronds,* and use as seasoning or brew like tea to make soup.   ALL YEAR

LAVER   **Fronds**
Porphyra spp.
Frond very thin, filmy, with satiny sheen; red, purple, or purple-brown; simple or variously lobed or cleft, with strongly ruffled margins; up to 1 ft. (30 cm) long. **Where found:** Rocks, piers, near low-tide level; often washed ashore. Throughout.
Use: Cooked vegetable, soup, nibble. Boil until tender, or dry and eat like Dulse (above).   ALL YEAR

SEA LETTUCE   **Fronds**
Ulva lactuca
Similar to Laver (above), but bright green. **Where found:** Rocks exposed to wave action. Throughout.
Use: Salad, seasoning. Add finely chopped to salads, or use dried and powdered as a saltlike seasoning.   ALL YEAR

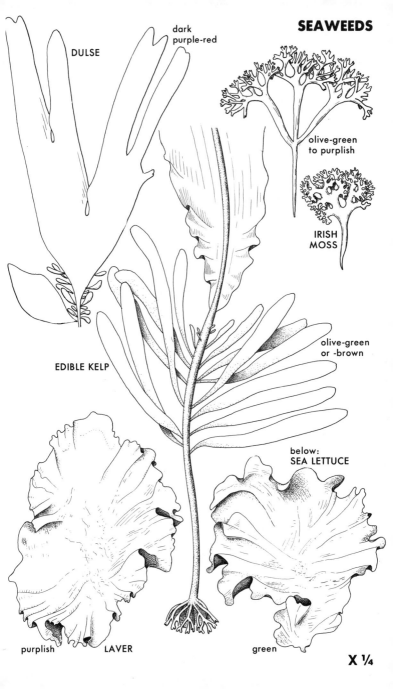

SEAWEEDS

DULSE

dark purple-red

olive-green to purplish

IRISH MOSS

olive-green or -brown

EDIBLE KELP

below: SEA LETTUCE

purplish LAVER

green

X ¼

LICHENS

Warning: The species below contain bitter rock-dissolving acids that are somewhat cathartic. Before using, soak for 2–3 hrs. in several changes of cold water (if available, add 1 tsp. of baking soda to each pot of water).

ROCK TRIPE   **Entire plant**
Umbilicaria spp. and *Gyrophora* spp.
Common, gray to olive-brown (darker beneath) lichens attached *blisterlike* to rocks. Circular or nearly so, 1–8 in. (2.5–20 cm) wide, attached at center; leathery when wet, brittle when dry. Shown is SMOOTH ROCK TRIPE, *U. mammulata.* **Where found:** Rocks, dry or moist open woods. Arctic south to n. U.S., in mts. to Ga.
Use: Cooked vegetable. Simmer for 1 hr. or add to soups and stews.   ALL YEAR

REINDEER MOSS   **Entire plant**
Cladonia rangiferina
A common northern lichen growing in large irregular colonies. Stem and branches *antherlike,* round, hollow, *ashy- to silvery-gray* (not yellowish- or greenish-gray). Pliant when wet, brittle when dry. 2–4 in. (5–10 cm). **Where found:** Open ground or partial shade. Arctic south to n. U.S.; most abundant northward.
Use: Flour, soup, blancmange. Powder the dried plant, and mix with other flour, add to soup, or use like cornstarch to make blancmange.   ALL YEAR

ICELAND MOSS   **Entire plant**
Cetraria islandica
A common far-northern lichen forming large tufted olive-green to brown mats. Stems and branches *strap-shaped,* paper-thin, freely forking, with tiny uniform spines along edges; flattened when wet, curling *troughlike* when dry; often with red splotches. 1–3 in. (2.5–7.5 cm). **Where found:** Sandy soil, exposed places. Arctic south to mt. tops of n. U.S.
Use: Flour, soup, blancmange. See Reindeer Moss (above).
                                               ALL YEAR

236

LICHENS

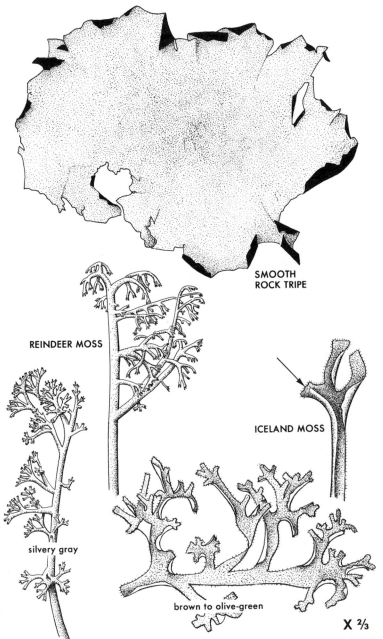

SMOOTH
ROCK TRIPE

REINDEER MOSS

ICELAND MOSS

silvery gray

brown to olive-green

X ⅔

MUSHROOMS

Warning: There are no foolproof methods for determining edible or poisonous mushrooms. Beginners should limit themselves to the four readily identifiable species below. If you wish to experiment with additional species, refer to a mushroom guide.

GIANT PUFFBALL **Fruiting body**
Calvatia gigantea
Fruiting body large, globular, smooth, growing directly from ground; white when young, dingy later. 8–15 in. (20–37.5 cm) across, sometimes considerably larger. **Where found:** Rich disturbed soil; open places, barnyards, pastures. Throughout. **Use:** Cooked vegetable. Excellent prepared like cultivated mushrooms. **Warning:** Make sure the interior flesh is *pure white;* bitter when yellowish. Cut open to be certain there is no rudimentary stem or gills. LATE SUMMER–FALL

SHAGGY MANE **Cap and stem**
Coprinus comatus
Like a closed erect umbrella when young, spreading with age. Cap white, with shinglelike brown or white scales, and white to pinkish gills; stem white, hollow, with a thin delicate ring. Colonial. 4–12 in. (10–30 cm). **Where found:** Roadsides, lawns, waste places. Throughout. **Use:** Cooked vegetable. Gather when young and use immediately. **Note:** Contains self-digestive enzymes that render older specimens to black fluid. FALL

COMMON MOREL **Cap and stem**
Morchella esculenta **Color pl. 14**
Cap pale tan to grayish, spongelike, oval or bluntly cone-shaped, fused to stem at lower end; deeply *pitted,* with ridges whitish. 2–4 in. (5–10 cm). BLACK MOREL, *M. angusticeps* (not shown) similar but with dark gray ridges. **Where found:** Moist woods, orchards, burned fields. Most of our area. **Use:** Cooked vegetable. Outstanding sautéed in butter. **Caution:** Do not confuse with false morels, *Gyromitra* spp. and *Verpa* spp., which have caps that hang skirtlike about the stem (not attached at lower end) and are convoluted like a brain (not pitted). Avoid morel-like mushrooms of the summer and fall; they are usually false morels. SPRING (usu. May)

SULPHUR SHELF, CHICKEN-OF-THE-WOODS
Polyporus sulphureus **Tender edges**
Color pl. 13 (only)
Forms large, many-leveled, *sulphur-yellow to orange brackets* on tree trunks or logs. Underside with *minute holes (pores).* **Where found:** Dead or injured deciduous trees. Throughout. **Use:** Cooked vegetable. The thinly sliced tender edges are excellent in stews or simmered for 30 min.

LATE SUMMER–FALL

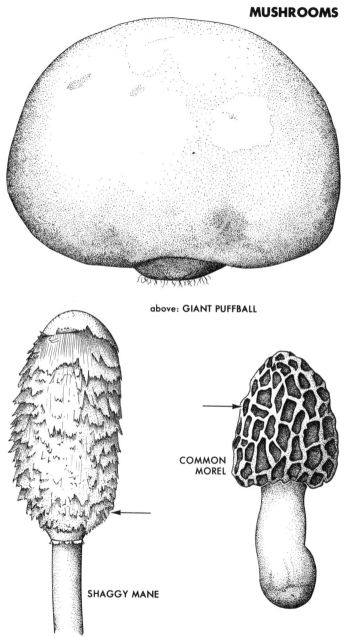

above: GIANT PUFFBALL

COMMON
MOREL

SHAGGY MANE

X ⅔

Finding Edible Plants

Where and When They Occur

Each plant is adapted to certain environmental conditions (temperature, moisture, sunlight, soil composition, and so forth) which are determined by climate, topography, geography, and associated plants. It will only occur where the combination of these conditions is suitable for its survival and growth. Such a place is called the plant's habitat. Before you go in search of any species, familiarize yourself with its habitat. This is the most efficient way of finding the plant. Usually several different species are adapted for living in the same habitat and form what is called a plant community. Since the presence of one member of the community frequently indicates the presence of others, it is often easiest to find a particular species by locating the dominant member of the plant community in which it lives, and then looking nearby.

Plant communities are not static, but are continually modifying their own environment. The accumulation of decaying organic material builds up the soil and increases its ability to support plant life. Gradually dry land is created out of wet land, and sterile soil becomes enriched with humus. For example, a pond will evolve into a swamp, and an open field will eventually become a forest. As each successive plant community gives way to the next, the habitat becomes more stable, and in time supports a climax community, such as a beech-hemlock forest. If there are no major disturbances (natural or manmade), the climax community will remain essentially unchanged.

Because succession is a dynamic process, habitats have a tendency to blend into one another, and plants will sometimes occur in unexpected places. In addition, minor variations in topography and soil types within a habitat can result in relic populations from a prior stage in the successional sequence, or forerunners of the next one.

The 14 habitats described in this section are the major ones found in our area. They are ordered in rough successional sequence (open water to dry land, bare earth to mature forest):

Seashores	p. 242	Open Fields	p. 262
Running Water	p. 245	Old Fields	p. 265
Still Water	p. 247	Thickets	p. 268
Northern Bogs	p. 249	Dry Open Woods	p. 272
Swamps	p. 251	Moist Woods	p. 275
Wet Meadows	p. 253	Spruce–Fir Woods	p. 281
Waste Ground	p. 256	Alpine Tundra	p. 283

Within each habitat, species are listed under the season(s) in which

241

they become available, and are grouped according to their food use. Abundant, frequently used species are in boldface. Symbols in the margins identify the food use groupings; these are defined in *Explanation of Symbols* (p. 15). Readers seeking edible plants available in the fall and early spring months should remember to look also under the heading Fall–Early Spring, as it includes many species not listed in either Fall or Spring.

General geographic designations for each species are indicated by the terms Far North, North, South, and Deep South in parentheses after the species name. Far North refers to arctic or subarctic species which are found at the extreme northern limits of our range (see Map of Area, p. XIV). North refers to species usually found from Missouri to Virginia and northward (a few extend along the higher Appalachians as far south as northern Georgia). The term South is used for species that occur primarily from Arkansas to North Carolina and southward. Deep South indicates species found along the sandy coastal plain of Georgia and the Gulf States. When no range designation is given, the species can be found in both the northern and southern parts of our range. Keep in mind that these geographic designations are crude, and you should refer to the main text to be sure a plant occurs in your particular area. **Note:** Some of the habitats described (moist woods, for example) embrace a broad range of conditions, and their species composition may vary considerably; in addition, biological pigeonholing is notoriously imprecise, so some species will not be consistently found in the habitats for which they are listed.

Seashores

The shores along the Atlantic and Gulf coasts yield an extraordinary array of wild foods: Seaweeds (p. 234) and shellfish from the intertidal zone; Sea-rocket (p. 100) and Beach Pea (pp. 122, 142) from the upper beach; Silverweed (p. 70), Glassworts (p. 146), Orache (p. 152), and Cattails (pp. 158, 230) from the saltmarshes. Behind barrier dunes, in areas protected from wind and spray, shrubs such as Beach Plum (p. 218), Bayberries (p. 206), Bearberry (pp. 36, 224), and Yaupon (p. 192) mark the transition to upland woods and thickets. **Note:** Dune communities tend to be in a state of delicate balance; overuse can lead to loss of ground cover and serious dune erosion.

Spring

Cattails, pp. 158, 230

Seaside Plantain (north),   Sea-blite, p. 146
   p. 46

Cattails, pp. 158, 230
**Saw Palmetto (deep
    south), p. 170**

Great Bulrush, p. 230
Laver, p. 234
Irish Moss (north), p. 234

 Irish Moss (north), p. 234

 Glassworts, p. 146

Cattails, pp. 158, 230

Spurge-nettle (deep south),
    p. 32

Cattails, pp. 158, 230

Seaside Plantain (north), p.
    46
Glassworts, p. 146
Cattails, pp. 158, 230
**Saw Palmetto (deep
    south), p. 170**

Great Bulrush, p. 230
Laver, p. 234
Dulse (north), p. 234
Sea Lettuce, p. 234
Edible Kelp (north), p. 234

Dulse (north), p. 234
Sea Lettuce, p. 234

Edible Kelp (north), p. 234

 Yaupon (south), p. 192

Summer

 **Wrinkled Rose (north),
    pp. 106, 184**

Reed, p. 228

Wild Rice, p. 228

**Seaside Plantain (north),
    p. 46**
Marsh Mallow (north),
    p. 108

Sea-purslane, p. 110
Sea-blite, p. 146
Orache, p. 152

Sea-rocket, p. 100
**Beach Pea (north), pp.
    122, 142**
Yuccas, p. 170

**Saw Palmetto (deep
    south), p. 170**
Irish Moss (north), p. 234
Dulse (north), p. 234

Cattails, pp. 158, 230
Wild Rice, p. 228

Great Bulrush, p. 230

Wrinkled Rose (north), pp.
    106, 184
**Beach Plum (north),
    p. 218**

Bearberry (north), pp. 36,
    224

Wrinkled Rose (north),
   pp. 106, 184
Beach Plum (north), p.
   218

Irish Moss (north), p. 234

Marsh Mallow (north),
   p. 108

Glassworts, p. 146

Spurge-nettle (deep
   south), p. 32

Yuccas, pp. 20, 170
Seaside Plantain (north),
   p. 46
Sea-rocket, p. 100
Glassworts, p. 146
Cattails, pp. 158, 230

Saw Palmetto (deep
   south), p. 170
Laver, p. 234
Dulse (north), p. 234
Sea Lettuce, p. 234
Edible Kelp (north), p. 234

Ground Juniper (north),
   p. 168
Bayberries, p. 206

Dulse (north), p. 234
Sea Lettuce, p. 234
Edible Kelp (north), p. 234

Wrinkled Rose (north),
   pp. 106, 184

Yaupon (south), p. 192

Fall

Reed, p. 228

Saw Palmetto (deep
   south), p. 170

Irish Moss (north), p. 234
Laver, p. 234

Reed, p. 228

Great Bulrush, p. 230

Wrinkled Rose (north), pp.
   106, 184
Beach Plum (north),
   p. 218

Bearberry (north), pp. 36,
   224

Wrinkled Rose (north),
   pp. 106, 184
Beach Plum (north),
   p. 218

Irish Moss (north), p. 234

Glassworts, p. 146

Spurge-nettle (deep south),
   p. 32

Glassworts, p. 146
Saw Palmetto (deep
   south), p. 170
Dulse (north), p. 234

Edible Kelp (north), p. 234
Laver, p. 234
Sea Lettuce, p. 234

Common Juniper (north),
   p. 168
Bayberries, p. 206

Dulse (north), p. 234
Edible Kelp (north), p. 234
Sea Lettuce, p. 234

Wrinkled Rose (north),
   pp. 106, 184

Yaupon (south), p. 192

Fall–Early Spring

Marsh Mallow (north),
   p. 108

Silverweed (north), p. 70
Marsh Mallow (north),
   p. 108
Saw Palmetto (deep
   south), p. 170

Irish Moss (north), p. 234
Laver, p. 234

Cattails, pp. 158, 230

Reed, p. 228

Irish Moss (north), p. 234

Great Bulrush, p. 230

Spurge-nettle (deep south),
   p. 32

Great Bulrush, p. 230

Saw Palmetto (deep
   south), p. 170
Dulse (north), p. 234

Edible Kelp (north), p. 234
Laver, p. 234
Sea Lettuce, p. 234

Dulse (north), p. 234
Edible Kelp (north), p. 234

Laver, p. 234
Sea Lettuce, p. 234

Yaupon (south), p. 192

Running Water (Streams, Springs)

Springs and small upland streams are dynamic, highly unstable
habitats whose bottoms are continually swept clean by rushing
water. Plants are limited to the margins, growing either in shallow
backwaters or in the wet disturbed soil along the shore.

The species listed below include only those edible plants that are
actually found growing at the very edge of running water. Not
included are field and woodland species such as Common Elder-

berry (pp. 18, 172), Bee-balm (p. 118), and Ostrich Fern (p. 232), which seem to prefer the rich alluvial soils found along the raised banks of streams; these are listed under Swamps (p. 251), Wet Meadows (p. 253), Thickets (p. 267), and Moist Woods (p. 274). Plants found along the placid margins of lowland streams and rivers are usually more typical of ponds and marshes, and are listed under Still Water (p. 247).

Spring

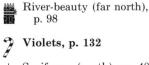 River-beauty (far north),
    p. 98

Violets, p. 132

 Saxifrages (north), pp. 48,
    146
Marsh-marigold (north),
    p. 70
Jewelweeds, pp. 78, 92

River-beauty (far north),
    p. 98
Violets, p. 132
Skunk Cabbage (north),
    p. 156

Marsh-marigold (north),
    p. 70

Pennsylvania Bitter-
    cress, p. 28
Mountain Watercress, p. 28
Watercress, p. 28
Spring Cress, p. 28
Cuckoo-flower (north), pp.
    28, 100

Saxifrages (north), pp. 48,
    146
American Brooklime
    (north), p. 128
Violets, p. 132

Spring Cress, p. 28

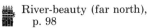 **Violets, p. 132**

Summer

 River-beauty (far north),
    p. 98

Virginia Meadow-beauty,
    p. 98

River-beauty (far north),
    p. 98

Watercress, p. 28
Virginia Meadow-beauty,
    p. 98

American Brooklime
    (north), p. 128

Mints, pp. 54, 118, 138

Mints, pp. 54, 118, 138

Fall–Early Spring

Water-parsnip, p. 42   Bugleweed, p. 54

Skunk Cabbage (north),
    p. 156

Bugleweed, p. 54

Watercress, p. 28   Bugleweed, p. 54

Still Water (Freshwater Ponds, Marshes)

The quiet or standing waters found in ponds, and placid backwaters along the margins of sluggish streams and rivers, support a very different community of plants than that associated with running water. Here the sediments are no longer swept away, but accumulate on the bottom. Plants are not limited to the margin; instead, they now extend out from the shore into deeper water. Marsh plants, such as Cattails (pp. 158, 230) and Wild Rice (p. 228) grow in shallow water. In deeper water are Pickerelweed (p. 136) and Arrowheads (p. 24). Floating-leaved plants, such as Water-lilies (p. 22) and Yellow Pond-lilies (p. 60), appear in much deeper water. Ponds and marshes are impermanent habitats. Sediments and decayed organic material will eventually fill them in, and they will become swampland.

A few of the species listed below, such as Shrubby Cinquefoil (pp. 70, 180), Silverweed (p. 70), and Hyssop Hedge-nettle (p. 104), are typically found on the exposed shores of some lakes and ponds. These wave-swept sandy or gravelly shores somewhat resemble the disturbed margins of fast-moving streams.

Spring

Cattails, pp. 158, 230   Large Cane (south), p. 228

Sweetflag, pp. 62, 156   Violets, p. 132

Water-lilies, p. 22   Water-shield, p. 96
American Lotus, p. 60   Violets, p. 132

Great Bulrush, p. 230   Cattails, pp. 158, 230
Water-hyacinth (deep
    south), p. 136

Cattails, pp. 158, 230

Cattails, pp. 158, 230

✗ Sweetflag, pp. 62, 156   **Cattails, pp. 158, 230**
  Water-shield, p. 96   **Great Bulrush, p. 230**
  Violets, p. 132

☕ **Violets, p. 132**

Summer

〈 Reed, p. 228

〈 **Pickerelweed, p. 136**   **Wild Rice, p. 228**
  Large Cane (south), p. 228

♨ Virginia Meadow-beauty,   Pickerelweed, p. 136
   p. 98

♨ Water-lilies, p. 22   Golden Club, p. 62
   Yellow Pond-Lilies, p. 60   Water-hyacinth (deep
   American Lotus, p. 60   south), p. 136

▦ **Yellow Pond-lilies, p. 60**   Cattails, pp. 158, 230
  American Lotus, p. 60   Large Cane (south), p. 228
  Golden Club, p. 62   **Wild Rice, p. 228**
  Pickerelweed, p. 136   Great Bulrush, p. 230

♠ American Lotus, p. 60

✗ Virginia Meadow-beauty,   Pickerelweed, p. 136
   p. 98   **Cattails, pp. 158, 230**

▮ **Mints, pp. 54, 118, 138**

☕ **Mints, pp. 54, 118, 138**
  Shrubby Cinquefoil
  (north), pp. 70, 180

Fall

〈 **Pickerelweed, p. 136**   Reed, p. 228

♨ Water-lilies, p. 22   American Lotus, p. 60
   Yellow Pond-lilies, 60   Golden Club, p. 62

▦ Wild Calla (north), p. 22   Golden Club, p. 62
  Water-lilies, p. 22   Pickerelweed, p. 136
  American Lotus, p. 60   Reed, p. 228
  Yellow Pond-lilies p. 60   Great Bulrush, p. 230

American Lotus, p. 60

Fall-Early Spring

Buckbean (north), p. 22   **Silverweed (north), p. 70**
Water-parsnip, p. 42   Hyssop Hedge-nettle, p. 104
Bugleweed, p. 54

Buckbean (north), p. 22   **Reed, p. 228**
Wild Calla (north), p. 22   **Cattails, pp. 158, 230**
Golden Club, p. 62   **Great Bulrush, p. 230**
Water-shield, p. 96

Bugleweed, p. 54
**Hyssop Hedge-nettle,
   p. 120**

Tuberous Water-lily   American Lotus, p. 60
   (north), p. 22   Water-shield, p. 96
Arrowheads, p. 24   **Great Bulrush, p. 230**
Yellow Pond-lilies, p. 60   Great Bur-reed, p. 230

Bugleweed, p. 54
**Hyssop Hedge-nettle,
   p. 120**

Northern Bogs

Scattered throughout the northern U.S. and extending as far south
as the Appalachians, bogs are wet sites typified by poor drainage,
acid water, low rates of decomposition, and large accumulations of
peat (partially decomposed organic material). Characteristic
plants are sphagnum moss (a group of gray-green to reddish
mosses that hold many times their weight in water), heath plants
(Blueberries, p. 220, Cranberries, pp. 102 and 222, Labrador-tea,
pp. 18 and 208, and others), and certain conifers (Black Spruce,
p. 168, Tamarack, p. 166, and Atlantic White Cedar, *Chamaecy-
paris thyoides*).

Bogs often form in water-filled depressions left behind by re-
treating glaciers. They begin with the introduction of sphagnum
moss and other early bog plants in among the stems of the marsh
plants growing along the shores. These early bog plants, tied
together by an extensive root system, form a floating mat which
slowly extends out over the surface of the water. The margins of
this mat remain fragile, but older portions often become thick
enough to support shrubs, people ("quaking bogs"), and even
trees. As peat accumulates in the tea-brown water and the mat
expands outwards, the water basin gradually grows smaller and

smaller. Eventually the bog becomes completely filled with peat, and the area evolves into forest.

Below are listed the edible plants typically associated with cold northern bogs. Plants that are normally found in areas of open water or in hardwood swamps surrounding bogs are listed under the headings "Still Water" (p. 247) and "Swamps" (p. 251). **Note:** Bogs are fragile habitats, and floating mats are easily damaged. Intrusions into boggy areas should be limited and as little disturbance created as possible.

Spring

Black Spruce, p. 168

Water Avens, p. 94

Tamarack, p. 166   Black Spruce, p. 168

Tamarack, p. 166   Black Spruce, p. 168

Labrador-tea, pp. 18, 208
**Creeping Snowberry, pp.
   36, 222**

Summer

Black Spruce, p. 168

Water Avens, p. 94

Arrow Arum, p. 156

Cloudberry (far north),   Northern Fly-honeysuckle,
   p. 32   pp. 78, 174
Creeping Snowberry, pp. 36,   **Blueberries, p. 220**
   222

Northern Fly-honeysuckle,   **Blueberries, p. 220**
   pp. 78, 174

Sweetgale, p. 206

Labrador-tea, pp. 18, 208   **Shrubby Cinquefoil, pp.**
Creeping Snowberry, pp.   **70, 180**
   36, 222   Sweetgale, p. 206

Fall

Black Spruce, p. 168

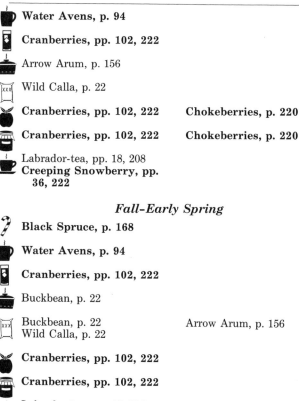

Water Avens, p. 94

Cranberries, pp. 102, 222

Arrow Arum, p. 156

Wild Calla, p. 22

Cranberries, pp. 102, 222   Chokeberries, p. 220

Cranberries, pp. 102, 222   Chokeberries, p. 220

Labrador-tea, pp. 18, 208
Creeping Snowberry, pp.
36, 222

Fall-Early Spring

Black Spruce, p. 168

Water Avens, p. 94

Cranberries, pp. 102, 222

Buckbean, p. 22

Buckbean, p. 22   Arrow Arum, p. 156
Wild Calla, p. 22

Cranberries, pp. 102, 222

Cranberries, pp. 102, 222

Labrador-tea, pp. 18, 208
Creeping Snowberry, pp.
36, 222

Swamps

Swamps are partially inundated woodlands. They represent the
final stage in the transformation of a wetland into dry land.
Swamps display considerable variation in their appearance and
species composition, ranging from cypress swamps in the deep
South to cedar swamps and alder thickets in the Northeast. The
formation of a swamp typically begins with the appearance of a
few shrubs along the margins of a marsh. As time passes, these
increase in number and shade out the original marsh plants.
Eventually most of the shrubs will give way to trees, and as

organic material accumulates, the swamp will slowly begin to revert to a moist wood. **Caution:** Before entering any swamp, be sure to familiarize yourself with the appearance of Poison Sumac (p. 186). Contact with any part of this plant will cause a severe rash.

Spring

Angelica (north), p. 40

Swamp Saxifrage (north), pp. 48, 146
Marsh-marigold (north), p. 70

Skunk Cabbage (north), p. 156

Angelica (north), p. 40
Tamarack (north), p. 166
Marsh-marigold (north), p. 70

Tamarack (north), p. 166

Cuckoo-flower (north), pp. 28, 100
Swamp Saxifrage (north), pp. 48, 146

American Brooklime (north), p. 128

Redbay (coastal south), p. 208

Common Spicebush, p. 208

Summer

Angelica (north), p. 40

Angelica (north), p. 40
Arrow Arum, p. 156

Juneberries, pp. 18, 220
Northern Fly-honeysuckle (north), pp. 78, 174
Wild-raisins, p. 178

Gooseberries and Currants, p. 212
Blueberries, p. 220

Juneberries, pp. 18, 220
Northern Fly-honeysuckle, (north), pp. 78, 174
Wild-raisins, p. 178

Gooseberries and Currants, p. 212
Blueberries, p. 220

American Brooklime (north), p. 128

Redbay (coastal south), p. 208
Sweetgale (north), p. 206
Common Wax-myrtle (south), p. 206

Common Spicebush, p. 208

Sweetgale (north), p. 206

Common Spicebush, p. 208

Fall

Arrow Arum, p. 156

Wild Calla (north), p. 22

Wild-raisins, p. 178

Chokeberries, p. 220

Wild-raisins, p. 178

Chokeberries, p. 220

Redbay (coastal south), p. 208
Common Wax-myrtle (south), p. 206

Common Spicebush, p. 208

Common Spicebush, p. 208

Fall–Early Spring

Water-parsnip, p. 42

Wild Calla (north), p. 22
Skunk Cabbage (north), p. 156

Jack-in-the-pulpit, p. 156
Arrow Arum, p. 156

Redbay (coastal south), p. 208

Common Spicebush, p. 208

Wet Meadows

Wet meadows are low-lying, predominantly grassy areas whose soil remains wet or moist throughout the year. In the prairie states, where they are called "low prairie," they occur as a natural step in the transition from marshland to open field ("dry prairie"). Elsewhere, however, they only occur where there has been some sort of disturbance. Normally the ground they occupy would be swampland, but the original trees and shrubs of the swamp have either been cut down or burned, and a lush growth of grasses and herbs has taken over. Most of the wet meadows in the eastern United States are manmade and were created for use as cow pastures. Usually they are kept open by periodic grazing or mowing, but in those instances where they are left undisturbed they eventually revert to swamp.

Spring

Angelica (north), p. 40
Sweetflag, pp. 62, 156

Violets, p. 132

Water Avens (north),
p. 94

Saxifrages (north), pp. 48, 146

Wild Onions, p. 114
Violets, pp. 132

Cow-parsnip (north), p. 40
Angelica (north), p. 40

Wild Onions, p. 114
Wild Hyacinth, p. 136

Wild Onions, p. 114

Spring Cress, p. 28
Cuckoo-flower (north), pp. 28, 100
Saxifrages (north), pp. 48, 146

Sweetflag, pp. 62, 156
Wild Onions, p. 114
American Brooklime (north), p. 128
Violets, p. 132

Spring Cress, p. 28

Wild Onions, p. 114

Violets, p. 132

Summer

Angelica (north), p. 40

Water Avens (north),
p. 94

Common Elderberry, pp.
18, 172

Virginia Meadow-beauty, p. 98

Angelica (north), p. 40
Canada Lily, p. 74

Turk's-cap Lily, p. 92
Wild Onions, p. 114

 Blue Vervain, p. 134

 **Common Elderberry, pp.
18, 172**

 **Common Elderberry, pp.
18, 172**

 **Common Elderberry, pp.
18, 172**

 Wild Onions, p. 114

European Great Burnet
(north), p. 94
Virginia Meadow-beauty,
p. 98

Wild Onions, p. 114
American Brooklime
(north), p. 128

Cow-parsnip (north), p. 40
Mints, pp. 54, 118, 138

Wild Onions, p. 114

 **Mints, pp. 54, 118, 138
Shrubby Cinquefoil
(north), pp. 70, 180**

Fall

**Water Avens (north),
p. 94**

Wild Onions, p. 114

Blue Vervain, p. 134

Wild Onions, p. 114

Wild Onions, p. 114

Wild Onions, p. 114

Fall–Early Spring

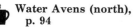 **Water Avens (north),
p. 94**

Cow-parsnip (north), p. 40
Water-parsnip, p. 42
Hyssop Hedge-nettle, p. 120

Bugleweed, p. 54
Wild Onions, p. 114
Woundwort (north), p. 120

Bugleweed, p. 54
Wild Onions, p. 114
Woundwort (north), p. 120

**Hyssop Hedge-nettle,
p. 120**

Bugleweed, p. 54
Wild Onions, p. 114
Woundwort (north), p. 120

**Hyssop Hedge-nettle, p.
120**

Wild Onions, p. 114

Waste Ground (Disturbed Soil)

Upland areas where the soil has been recently laid bare — fire
burns, lawns, gardens, cultivated fields, roadsides, railroad shoul-
ders, and vacant lots — are particularly rich collecting sites. An
extraordinarily large percentage of the annual and biennial plants
that quickly invade such waste places are edible. Many have
followed man's migration around the world (cosmopolitan weeds
such as Common Dandelion, p. 84, and Lamb's-quarters, p. 152)
and supply some of our finest wild foods.

The species that grow in any particular waste ground vary
according to the climate, and the chemical composition and moisture content of the soil. All waste places, however, have one
characteristic in common: within a few years all of the annual and
most biennial plants are crowded out by grasses and perennials
such as asters and goldenrods. The bare earth becomes an open
field.

Spring

**Japanese Knotweed,
p. 46**
Pokeweed, p. 46

Asparagus, p. 76
Day-lily, p. 92
Fireweed (north), p. 98

Coltsfoot (north), p. 84

Spiderworts, p. 130

Chufa, p. 230

Yellow Wood-sorrels, p. 72
Sorrels, pp. 116, 154

Chufa, p. 230

Peppergrasses, p. 26
Shepherd's Purse, p. 26

Field Pennycress, p. 26
Mallows, pp. 32, 108, 134

Bladder Campion (north), p. 34
Common Chickweed, p. 34
Mouse-ear Chickweed, p. 34
Caraway (north), p. 38
Common Plantain, p. 46
Pokeweed, p. 46
Cleavers, p. 50
Sweet Clovers, pp. 56, 80
Clovers, pp. 56, 80, 124
Chicory, pp. 58, 144
Wild Mustards, p. 64
Winter Cress, p. 64
Common Evening-primrose, p. 66
Comfrey, pp. 68, 110, 134
Common Dandelion, p. 84
Yellow Goat's-beard (north), p. 84

Sow-thistles, p. 86
Wild Lettuces, p. 86
Fireweed (north), p. 98
Storksbill, p. 104
Orpine (north), p. 112
Sorrels, pp. 116, 154
Lady's-thumb, p. 116
Burdocks, p. 126
Thistles, p. 126
Oyster-plant (north), p. 126
Dayflowers, p. 130
Spiderwort, p. 130
Corn-salad, p. 134
Blue Lettuces, p. 144
Nettles, p. 150
Amaranth, p. 154
Curled Dock, p. 154

Wild Mustards, p. 64
Winter Cress, p. 64
Common Dandelion, p. 84
Day-lily, p. 92

Wild Onions, p. 114
Burdocks, p. 126
Thistles, p. 126
Chufa, p. 230

Clovers, pp. 56, 80, 124

Chufa, p. 230

Common Dandelion, p. 84

Japanese Knotweed, p. 46

Pokeweed, p. 46

Wild Onions, p. 114

Spurge-nettle (deep south), p. 32

Shepherd's Purse, p. 26
Field Pennycress, p. 26
Peppergrasses, p. 26
Horseradish, p. 26
Common Chickweed, p. 34

Caraway (north), p. 38
Common Plantain, p. 46
Cleavers, p. 50
Sweet Clovers, pp. 56, 80
Clovers, pp. 56, 80, 124

✕ Chicory, pp. 58, 144
Ox-eye Daisy, p. 58
Wild Mustards, p. 64
Winter Cress, p. 64
Common Evening-primrose,
   p. 66
Yellow Wood-sorrels,
   p. 72
Common Dandelion,
   p. 84
Yellow Goat's-beard
   (north), p. 84
Sow-thistles, p. 86
Wild Lettuces, p. 86
Day-lily, p. 92
Storksbill, p. 104

Orpine (north), p. 112
Wild Onions, p. 114
Sorrels, pp. 116, 154
Lady's-thumb, p. 116
Alfalfa, pp. 124, 142
Burdocks, p. 126
Thistles, p. 126
Oyster-plant (north),
   p. 126
Dayflowers, p. 130
Spiderwort, p. 130
Corn-salad, p. 134
Blue Lettuces, p. 144
Amaranth, p. 154
Curled Dock, p. 154
Chufa, p. 230

Horseradish, p. 26
Sweet Clovers, pp. 56, 80
Coltsfoot (north), p. 84

Common Tansy, p. 90
Wild Onions, p. 114

Clovers, pp. 56, 80, 124
Comfrey, pp. 68, 110, 134
Coltsfoot (north), p. 84

Alfalfa, pp. 124, 142
Gill-over-the-ground, p. 140
Nettles, p. 150

Summer

Lesser Broomrape, p. 128

Sicklepod, p. 82

Coltsfoot (north), p. 84
Great Burdock, p. 126

Spiderwort, p. 130

Common Sunflower (west),
   p. 88

Cleavers, p. 50
Scotch Broom, pp. 82, 182
Common Sunflower (west),
   p. 88

Chufa, p. 230

Yellow Wood-sorrels, p. 72
Passion-flower (south),
   p. 78

Sorrels, pp. 116, 154
Chufa, p. 230

Shepherd's Purse, p. 26
Field Pennycress, p. 26
Mallows, pp. 32, 108, 134
Common Chickweed,
   p. 34

Mouse-ear Chickweed, p. 34
Clovers, pp. 56, 80, 124
Galinsoga, p. 58
Pilewort, p. 58
Purslane, p. 72

Sorrels, pp. 116, 154
Thistles, p. 126
Dayflowers, p. 130
Nettles, p. 150

Lamb's-quarters, p. 152
Strawberry-blite (north),
  p. 152
Amaranth, p. 154

Wild Mustards, p. 64
Wild Lettuces, p. 86
Day-lily, p. 92
Wild Onions, p. 114
Burdocks, p. 126
Thistles, p. 126

Creeping Bellflower (north),
  p. 128
Blue Lettuces, p. 144
Chufa, p. 230
Giant Puffball, p. 238

Clovers, pp. 56, 80, 124
Purslane, p. 72
Black Medick, p. 80
**Common Sunflower
  (west), p. 88**

Blue Vervain, p. 134
Chufa, p. 230

**Common Dandelion,
  p. 84**

Day-lily, p. 92

**Common Strawberry,
  p. 30**
Ground-cherries, p. 68
**Passion-flower (south),
  p. 94**

Strawberry-blite (north),
  p. 152

**Common Strawberry,
  p. 30**
Ground-cherries, p. 68

**Passion-flower (south),
  p. 94**

**Common Sunflower
  (west), p. 88**

Wild Mustards, p. 64
Purslane, p. 72

Scotch Broom, pp. 82, 182
Wild Onions, p. 114

Spurge-nettle (deep south),
  p. 32

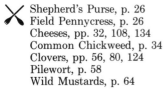

Shepherd's Purse, p. 26
Field Pennycress, p. 26
Cheeses, pp. 32, 108, 134
Common Chickweed, p. 34
Clovers, pp. 56, 80, 124
Pilewort, p. 58
Wild Mustards, p. 64

Purslane, p. 72
**Yellow Wood-sorrels,
  p. 72**
Wild Onions, p. 114
Sorrels, pp. 116, 154
Alfalfa, pp. 124, 142
Great Burdock, p. 126

Thistles, p. 126
Creeping Bellflower (north),
   p. 128

Dayflowers, p. 130
Amaranth, p. 154
Chufa, p. 230

Shepherd's Purse, p. 26
Field Pennycress, p. 26
Peppergrasses, p. 26
Horseradish, p. 26
Caraway (north), p. 38
Sweet Clovers, pp. 56, 80

Wild Mustards, p. 64
Coltsfoot (north), p. 84
Common Tansy, p. 90
Day-lily, p. 92
Wild Onions, p. 114
Basil (north), p. 120

Common Strawberry, p. 30
Yarrow, p. 38
Catnip, pp. 54, 140
Clovers, pp. 56, 80, 124
Wild Chamomile (north),
   p. 58
Common Mullein, p. 72
Coltsfoot (north), p. 84
Pineapple-weed (north),
p. 90

Fireweed (north), p. 98
Basil (north), p. 120
Alfalfa, pp. 124, 142
American Pennyroyal,
   p. 140
Gill-over-the-ground, p. 140
Nettles, p. 150
Mexican-tea, p. 152
Jerusalem-oak, p. 152

Fall

Common Sunflower (west),
   p. 88

Lamb's-quarters, p. 152

Chufa, p. 230

Common Sunflower (west),
   p. 88

Yellow Wood-sorrels, p. 72
Passion-flower (south),
p. 94

Chufa, p. 230

Common Chickweed,
p. 34
Mouse-ear Chickweed, p. 34

Thistles, p. 126
Corn-salad, p. 134
Amaranth, p. 154

Day-lily, p. 92
Wild Onions, p. 114
Thistles, p. 126
Creeping Bellflower (north),
   p. 128

Chufa, p. 230
Giant Puffball, p. 238
Shaggy Mane, p. 238

Black Medick, p. 80
Common Sunflower
(west), p. 88
Blue Vervain, p. 134

Lamb's-quarters, p. 152
Amaranth, p. 154
Chufa, p. 230

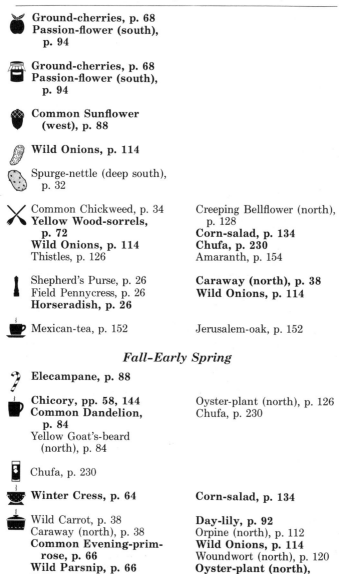

Ground-cherries, p. 68
Passion-flower (south),
     p. 94

Ground-cherries, p. 68
Passion-flower (south),
     p. 94

**Common Sunflower
(west), p. 88**

Wild Onions, p. 114

Spurge-nettle (deep south),
     p. 32

Common Chickweed, p. 34   Creeping Bellflower (north),
Yellow Wood-sorrels,   p. 128
     p. 72   **Corn-salad, p. 134**
Wild Onions, p. 114   **Chufa, p. 230**
Thistles, p. 126   Amaranth, p. 154

Shepherd's Purse, p. 26   **Caraway (north), p. 38**
Field Pennycress, p. 26   **Wild Onions, p. 114**
Horseradish, p. 26

Mexican-tea, p. 152   Jerusalem-oak, p. 152

Fall–Early Spring

Elecampane, p. 88

Chicory, pp. 58, 144   Oyster-plant (north), p. 126
Common Dandelion,   Chufa, p. 230
     p. 84
Yellow Goat's-beard
     (north), p. 84

Chufa, p. 230

Winter Cress, p. 64   **Corn-salad, p. 134**

Wild Carrot, p. 38   **Day-lily, p. 92**
Caraway (north), p. 38   Orpine (north), p. 112
Common Evening-prim-   **Wild Onions, p. 114**
     rose, p. 66   Woundwort (north), p. 120
Wild Parsnip, p. 66   **Oyster-plant (north),**
Yellow Goat's-beard   **p. 126**
     (north), p. 84   **Chufa, p. 230**

Chufa, p. 230

Jerusalem Artichoke,   **Wild Onions, p. 114**
    p. 88   Woundwort (north), p. 120
Orpine (north), p. 112

Wild Potato-vine, p. 20   **Jerusalem Artichoke,**
Spurge-nettle (deep south),   **p. 88**
    p. 32

Winter Cress, p. 64   **Wild Onions, p. 114**
Jerusalem Artichoke,   Woundwort (north), p. 120
    p. 88   **Corn-salad, p. 134**
Day-lily, p. 92   **Chufa, p. 230**

Wild Onions, p. 114   **Horseradish, p. 26**

Open Fields

The dominant feature of an open field is, of course, its grasses.
Along with these are large numbers of showy perennials that often
give the habitat the appearance of a wildflower garden. There are
no annual herbs, such as the ones found in disturbed soil, and very
few biennials. There are also no woody plants. The species
makeup and general appearance of open fields are subject to a
great deal of variation and depend largely on the quality and
moisture content of the soil. They range from lush hayfields to dry
sterile fields with a few native grasses and drought-resistant herbs.

Fields are dynamic, rapidly changing habitats and unless they
are kept open by burning or mowing, will revert back to forest.
Trees and shrubs will usually begin to invade the field within three
to five years after it is left fallow. When they do, the habitat is
more appropriately termed an old field (see p. 265).

Spring

Common Milkweed,
    p. 112

Spiderworts, p. 130

Sorrels, pp. 116, 154

Shepherd's Purse, p. 26   **Yellow Goat's-beard**
Field Pennycress, p. 26   **(north), p. 84**
Cow-cress, p. 26   **Sow-thistles, p. 86**
Mouse-ear Chickweed, p. 34   **Wild Lettuces, p. 86**
Clovers, pp. 56, 124   Storksbill, p. 104

Musk Mallow (north),
  p. 108
Orpine (north), p. 112
Common Milkweed,
  p. 112
Wild Onions, p. 114

Sorrels, pp. 116, 154
Thistles, p. 126
Oyster-plant (north),
  p. 126
Spiderwort, p. 130

Wild Onions, p. 114

Thistles, p. 126

Clovers, pp. 56, 124

Wild Onions, p. 114

Shepherd's Purse, p. 26
Field Pennycress, p. 26
Cow-cress, p. 26
Clovers, pp. 56, 124
Ox-eye Daisy, p. 58
Yellow Goat's-beard
  (north), p. 84
Sow-thistles, p. 86
Wild Lettuces, p. 86

Storksbill, p. 104
Wild Onions, p. 114
Orpine (north), p. 112
Sorrels, pp. 116, 154
Thistles, p. 126
Oyster-plant (north),
  p. 126
Spiderwort, p. 130

Wild Onions, p. 114

Clovers, pp. 56, 124

Summer

Lesser Broomrape, p. 128

Compass-plant (west), p. 88

Spiderwort, p. 130

Common Sunflower (west),
  p. 88

Common Sunflower (west),
  p. 88

Sorrels, pp. 116, 154

Shepherd's Purse, p. 26
Field Pennycress, p. 26
Mouse-ear Chickweed, p. 34
Clovers, pp. 56, 124

Musk Mallow (north),
  p. 108
Sorrels, pp. 116, 154
Thistles, p. 126

Wild Lettuces, p. 86
Wild Onions, p. 114
Common Milkweed,
  p. 112

Thistles, p. 126

Common Sunflower (west), p. 88

Clovers, pp. 56, 124

Common Milkweed, p. 112

Common Strawberry, p. 30

Ground-cherries, p. 68

Common Strawberry, p. 30

Ground-cherries, p. 68

Common Sunflower (west), p. 88

Wild Onions, p. 114

Shepherd's Purse, p. 26
Field Pennycress, p. 26
Clovers, pp. 56, 124

Wild Onions, p. 114
Sorrels, pp. 116, 154
Thistles, p. 126

Shepherd's Purse, p. 26
Field Pennycress, p. 26

Cow-cress, p. 26
Wild Onions, p. 114

Common Strawberry, p. 30
Yarrow, p. 38
Mountain-mints, p. 54
Clovers, pp. 56, 124

Common Mullein, p. 72
Wild Bergamot, p. 118
Blue Giant Hyssop (west), p. 140

Fall

Common Sunflower (west), p. 88

Common Sunflower (west), p. 88

Mouse-ear Chickweed, p. 34

Thistles, p. 126

Wild Onions, p. 114

Thistles, p. 126

Common Sunflower (west), p. 88

Ground-cherries, p. 68

Ground-cherries, p. 68

Common Sunflower (west), p. 88

Wild Onions, p. 114

Wild Onions, p. 114   Thistles, p. 126

Shepherd's Purse, p. 26   **Wild Onions, p. 114**
Field Pennycress, p. 26

Fall–Early Spring

Elecampane, p. 88

Yellow Goat's-beard   Oyster-plant (north), p. 126
  (north), p. 84

Wild Carrot, p. 38   **Wild Onions, p. 114**
Yellow Goat's-beard   **Oyster-plant (north),**
  (north), p. 84   **p. 126**
Orpine (north), p. 112

Orpine (north), p. 112   **Wild Onions, p. 114**

Wild Onions, P. 114

Wild Onions, p. 114

Old Fields

Old fields share some of the characteristics of both an open field
and a thicket. Grasses and perennial wildflowers are still abundant, but woody plants such as Roses (pp. 106, 184), Sumacs
(p. 186), Brambles (pp. 30, 184) Sassafras (p. 210), and Northern
Bayberry (p. 206) have begun to invade the field in significant
numbers. This invasion is most advanced along those margins of a
field adjacent to woods.

As trees and shrubs appear in ever increasing numbers, they
begin to coalesce into distinct thicketlike clumps and start shading
out the original grasses and herbs. At this late stage in the development of an old field, some fruit-bearing shrubs and trees such as
Common Barberry (p. 192) and Hawthorns (p. 216) produce their
heaviest crops of fruit. These same species are, of course, still
present when an old field has reverted to a thicket, but at that
stage overcrowding and increased competition for light usually
result in smaller yields. Eventually, the various thicketlike clumps
will begin to merge and, often tied together by vines, will become
impenetrable walls of vegetation. The field is then a thicket.

Spring

 Asparagus, p. 76
Common Milkweed,
p. 112

Bracken, p. 232

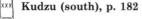 **Common Milkweed,**
p. 112

Prickly-pear, p. 60

Kudzu (south), p. 182

Black Locust, p. 184

Spurge-nettle (deep south),
p. 32

Blackberries, pp. 30, 184

Bracken, p. 232

Sassafras, p. 210

Sassafras, p. 210

Summer

Roses, pp. 106, 184

Brambles, pp. 30, 184
Passion-flower (south),
p. 94

Highbush-cranberry
(north), p. 178
Sumacs, p. 186

Prickly-pear, p. 60
Common Milkweed,
p. 112

Yuccas, p. 170

Prickly-pear, p. 60

Kudzu (south), p. 182

Common Milkweed, p. 112

Common Strawberry,
p. 30
Brambles, pp. 30, 184
Prickly-pear, p. 60
Ground-cherries, p. 68

Passion-flower (south),
p. 94
Roses, pp. 106, 184
Highbush-cranberry
(north), p. 178

Common Strawberry,
p. 30
Brambles, pp. 30, 184
Ground-cherries, p. 68
Passion-flower (south),
p. 94

Roses, pp. 106, 184
Highbush-cranberry
(north), p. 178

Spurge-nettle (deep south),
  p. 32

Yuccas, pp. 20, 170

Ground Juniper (north),   **Northern Bayberry**
  p. 168   **(north), p. 206**
Sassafras, p. 210

Common Strawberry, p. 30   Bergamots, p. 118
Brambles, pp. 30, 184   **Persimmon, p. 194**
Mountain-mints, p. 54   Sweetfern (north), p. 206
Common Mullein, p. 72   **Sassafras, p. 210**
Roses, pp. 106, 184

Fall

Passion-flower (south),   **Common Barberry**
  p. 94   **(north), p. 192**
Highbush-cranberry   **Domestic Apple, p. 216**
  (north), p. 178

Prickly-pear, p. 60   Kudzu (south), p. 182

Prickly-pear, p. 60   **Common Barberry**
Ground-cherries, p. 68   **(north), p. 192**
Passion-flower (south),   **Persimmon, p. 194**
  p. 94   **Domestic Apple, p. 216**
Roses, pp. 106, 184   **Domestic Pear, p. 216**
Highbush-cranberry
  (north), p. 178

Ground-cherries, p. 68   **Common Barberry**
Passion-flower (south),   **(north), p. 192**
  p. 94   **Persimmon, p. 194**
Roses, pp. 106, 184   **Hawthorns, p. 216**
Highbush-cranberry   **Crabapples, p. 216**
  (north), p. 178   **Domestic Apple, p. 216**

Wild Potato-vine, p. 20   Spurge-nettle (deep south),
                                        p. 32

Ground Juniper (north),   **Northern Bayberry**
  p. 168   **(north), p. 206**

Roses, pp. 106, 184   Hawthorns, p. 216
Sassafras, p. 210

Fall-Early Spring

 Kudzu (south), p. 182

Persimmon, p. 194

Wild Potato-vine, p. 20
Spurge-nettle (deep south),
    p. 32

Sassafras, p. 210

Thickets

A thicket is an impenetrable tangle of shrubs, vines, and small trees. Competition in a thicket is very intense, with each plant trying to overtop the other in an effort to get the most sunlight. Except for a few of the taller and more competitive herbaceous plants, such as Asparagus (p. 76) and Jerusalem Artichoke (p. 88), virtually all the grasses and herbs from the old field have long since been shaded out. Even these last few will quickly disappear and leave only a relatively minor and unimportant group of shade-tolerant herbs and grasses. During the early part of their existence, when there is plenty of sunlight, thickets produce abundant crops of fruit. Later, however, these heavy yields diminish as trees like Sassafras (p. 210), Red Maple (p. 176), and Black Birch (p. 200) begin to overshadow and dominate the thicket. Eventually, the smaller fruit-bearing trees and shrubs will disappear altogether, and what originally may have begun as plowed field will have become forest.

Spring

Asparagus, p. 76   Greenbriers, pp. 148, 196
Carrion-flower, p. 148

Greenbriers, pp. 148, 196

Cleavers, p. 50   Blue Lettuces, p. 144
Wild Lettuces, p. 86   **Greenbriers, pp. 148, 196**

Cow-parsnip (north), p. 40   **Mulberries, p. 210**
Redbud, pp. 122, 192

 Carrion-flower, p. 148   **Kudzu (south), p. 182**
Greenbriers, pp. 148, 196

Wisteria, p. 180 **Black Locust, p. 184**

Carrion-flower, p. 148 Greenbriers, pp. 148, 196

Redbud, pp. 122, 192

Groundnut, pp. 122, 160

Blackberries, pp. 30, 184 **Redbud, pp. 122, 192**
Cleavers, p. 50 Blue Lettuces, p. 144
Wild Lettuces, p. 86 **Greenbriers, pp. 148, 196**

Sassafras, p. 210

Yaupon (south), p. 192 Sassafras, p. 210

Summer

Greenbriers, pp. 148, 196

Roses, pp. 106, 184 Hazelnuts, p. 200

Cleavers, p. 50

Common Elderberry, pp. 18, 172 Greenbriers, pp. 148, 196
Brambles, pp. 30, 184 **Highbush-cranberry (north), p. 178**
Passion-flower (south), p. 94 **Sumacs, p. 186**
Red Mulberry, p. 210

Greenbriers, pp. 148, 196 Wild Grapes, p. 198

Wild Lettuces, p. 86 Wild Bean, p. 124
Redbud, pp. 122, 192 Blue Lettuces, p. 144

Carrion-flower, p. 148 **Kudzu (south), p. 182**
Greenbriers, pp. 148, 196 **Hazelnuts, p. 200**

Common Elderberry, pp. 18, 172

Common Elderberry, pp. 18, 172 Purple-flowering Raspberry (north) pp. 106, 212
Juneberries, pp. 18, 220 **Wild-raisins, p. 178**
Brambles, pp. 30, 184 **Highbush-cranberry (north), p. 178**
Passion-flower (south), p. 94 **Mulberries, p. 210**
Roses, pp. 106, 184

Gooseberries and Currants, p. 212
Wild Cherries, p. 218

Wild Plums, p. 218
Blueberries, p. 220

Common Elderberry, pp. 18, 172
Juneberries, pp. 18, 220
Brambles, pp. 30, 184
Passion-flower (south), p. 94
Roses, pp. 106, 184
Carrion-flower, p. 148
Greenbriers, pp. 148, 196

Wild-raisins, p. 178
Highbush-cranberry (north), p. 178
Red Mulberry, p. 210
Gooseberries and Currants, p. 212
Wild Cherries, p. 218
Wild Plums, p. 218
Blueberries, p. 220

Hazelnuts, p. 220

Groundnut, pp. 122, 160

Bastard-toadflax, p. 48

Greenbriers, pp. 148, 196

Cow-parsnip (north), p. 40
Basil (north), p. 120

Bayberries, p. 206
Sassafras, p. 210

Brambles, pp. 30, 184
Mountain-mints, p. 54
Roses, pp. 106, 184
Bergamots, p. 118
Basil (north), p. 120

Downy Wood-mint, p. 138
Blue Giant Hyssop (west), p. 140
Yaupon (south), p. 192
Sassafras, p. 210

Fall

Hazelnuts, p. 200
Eastern Chinquapin, p. 202

Feverwort, p. 160

Passion-flower (south), p. 94
Greenbriers, pp. 148, 196
Highbush-cranberry (north), p. 194

Common Barberry (north), p. 192
Wild Grapes, p. 198
Domestic Apple, p. 216

Wild Bean, p. 124

Carrion-flower, p. 148
Greenbriers, pp. 148, 196
Kudzu (south), p. 182

Hazelnuts, p. 200
Eastern Chinquapin, p. 202

Passion-flower (south), p. 94

Roses, pp. 106, 184
Wild-raisins, p. 178

Highbush-cranberry
  (north), p. 178
Common Barberry
  (north), p. 192
Wild Grapes, p. 198

Domestic Apple, p. 216
Domestic Pear, p. 216
Wild Cherries, p. 218
Wild Plums, p. 218
Chokeberries, p. 220

Passion-flower (south),
  p. 94
Roses, pp. 106, 184
Carrion-flower, p. 148
Greenbriers, pp. 148, 196
Wild-raisins, p. 178
Highbush-cranberry
  (north), p. 178
Common Barberry
  (north), p. 192

Wild Grapes, p. 198
Hawthorns, p. 216
Domestic Apple, p. 216
Crabapples, p. 216
Wild Cherries, p. 218
Wild Plums, p. 218
Chokeberries, p. 220

Hazelnuts, p. 200

Eastern Chinquapin,
  p. 202

Groundnut, pp. 122, 160

Bayberries, p. 206

Roses, pp. 106, 184
Yaupon (south), p. 192

Sassafras, p. 210
Hawthorns, p. 216

Fall–Early Spring

Greenbriers, pp. 148, 196

Cow-parsnip (north), p. 40

Hog-peanut, p. 124

Carrion-flower, p. 148
Greenbriers, pp. 148, 196

Kudzu (south), p. 182

Carrion-flower, p. 148

Greenbriers, pp. 148, 196

Jerusalem Artichoke,
  p. 88

Wild Potato-vine, p. 20
Jerusalem Artichoke,
  p. 88

Groundnut, pp. 122, 160

Jerusalem Artichoke,
  p. 88

☕ **Yaupon (south), p. 192**   **Sassafras, p. 210**

Dry Open Woods

Dry open woods form in sandy, well-drained soils typical of the outer coastal plain, extending in a wide band from the Southeast to a narrower strip in the North, and around the southern shores of the Great Lakes. They are also found along the tops of ridges where the soil is shallow, subject to rapid runoff, and exposed to the drying effects of wind and sun. Woods in such areas tend to be more open (more breaks in the canopy) than those defined as moist woods. Forest floors are sunlit. The typical trees are scrub pines and oaks. Growing beneath them are low shrubs such as Northern Bayberry (p. 206) and Blueberries (p. 220), grasses, and drought-resistant herbs such as Bracken (p. 232), Prickly-pear (p. 60), and Mountain-mints (p. 54).

Dry open woods will naturally evolve into moist woods, so in some cases the dividing line is hazy. Pines are replaced by taller oaks, which become the dominant species. Such woods may have characteristics of both types of forests. See also Moist Woods (p. 275).

Spring

Greenbriers, pp. 148, 196   Bracken, p. 232

Pines, p. 166

Greenbriers, pp. 148, 196

Buffalo Clover, p. 56   **Greenbriers, pp. 148, 196**
Hairy Lettuce, p. 86
Large-leaved Aster (north),
   p. 144

Prickly-pear, p. 60   **Saw Palmetto (deep**
Pines, p. 166   **south), p. 170**
Rock Tripe (north), p. 236

Buffalo Clover (north),   Pines, p. 166
   p. 56   Reindeer Moss (north),
Greenbriers, pp. 148, 196   p. 236

Black Locust, p. 184

Greenbriers, pp. 148, 196

Spurge-nettle (deep south),
   p. 32

Trailing Arbutus, pp. 36,   Buffalo Clover (north),
   102, 224   p. 56

Hairy Lettuce, p. 86
Greenbriers, pp. 148, 196
**Saw Palmetto (deep
south), p. 170**

Bracken, p. 232

 Hickories, p. 190

Wintergreen (north), pp.
36, 224
Buffalo Clover (north),
p. 56

Pines, p. 166

Summer

Greenbriers, pp. 148, 196

 Hazelnuts, p. 200

Scotch Broom, pp. 82, 182

Greenbriers, pp. 148, 196
Canada Buffaloberry (far
north), p. 174

 Buffalo Clover (north),
p. 56

Greenbriers, pp. 148, 196

Prickly-pear, p. 60
Hairy Lettuce, p. 86
Wild Bean, p. 124
**Saw Palmetto (deep
south), p. 170**

Yuccas, p. 170
Rock Tripe (north), p. 236
Sulphur Shelf, p. 238

Buffalo Clover (north),
p. 56
Prickly-pear, p. 60
Greenbriers, pp. 148, 196

Hazelnuts, p. 200
Reindeer Moss (north),
p. 236

Juneberries, pp. 18, 220
Bearberry (north), pp. 36,
224
Prickly-pear, p. 60

Canada Buffaloberry (far
north), p. 174
Wild-raisins, p. 178
Blueberries, p. 220

Juneberries, pp. 18, 220
Greenbriers, pp. 148, 196
Canada Buffaloberry (far
north), p. 174

Wild-raisins, p. 178
Blueberries, p. 220

Hazelnuts, p. 200

Scotch Broom, pp. 82, 182

Spurge-nettle (deep south),
p. 32

Yuccas, pp. 20, 170
Bastard-toadflax, p. 48
Buffalo Clover (north),
p. 56

Greenbriers, pp. 148, 196
**Saw Palmetto (deep
south), p. 170**

**Northern Bayberry
(north), p. 206**

**New Jersey Tea, pp. 18,
192**
**Wintergreen (north), pp.
36, 224**
Mountain-mints, p. 54
**Buffalo Clover (north),
p. 56**
Whorled Loosestrife, p. 72
Horsemint, p. 78
Sweet Goldenrod, p. 90

Dittany, p. 120
Downy Wood-mint, p. 138
American Pennyroyal,
p. 140
Blue Giant Hyssop (west),
p. 140
Pines, p. 166
Persimmon, p. 194
Sweetfern (north), p. 206

Fall

Hickories, p. 190
Hazelnuts, p. 200
Chestnut, p. 202

**Eastern Chinquapin,
p. 202**
Oaks, p. 204

Feverwort, p. 160

Greenbriers, pp. 148, 196

Wild Bean, p. 124
**Saw Palmetto (deep
south), p. 170**

Rock Tripe (north), p. 236
Sulphur Shelf, p. 238

Prickly-pear, p. 60
Greenbriers, pp. 148, 196
Hickories, p. 190
Hazelnuts, p. 200
Chestnut, p. 202

**Eastern Chinquapin,
p. 202**
Oaks, p. 204
Reindeer Moss, p. 236

Bearberry (north), pp. 36,
224
Prickly-pear, p. 60

Wild-raisins, p. 178
Persimmon, p. 194

Greenbriers, pp. 148, 196

Wild-raisins, p. 178

Hickories, p. 190
Hazelnuts, p. 200
Chestnut, p. 202

Eastern Chinquapin,
  p. 202
Oaks, p. 204

Spurge-nettle (deep south),
  p. 32

Saw Palmetto (deep
  south), p. 170

Northern Bayberry
  (north), p. 206

Wintergreen (north), pp.
  36, 224

Pines, p. 166

Fall-Early Spring

Greenbriers, pp. 148, 196

Saw Palmetto (deep
  south), p. 170

Rock Tripe (north),
  p. 236

Greenbriers, pp. 148, 196
Reindeer Moss (north),
  p. 236

Wintergreen (north), pp.
  36, 224

Persimmon, p. 194

Greenbriers, pp. 148, 196

Persimmon, p. 194

Spurge-nettle (deep south),
  p. 32

Partridgeberry, pp. 36, 102,
  174
Wintergreen (north), pp. 36,
  224

Saw Palmetto (deep
  south), p. 170

Wintergreen (north), pp.
  36, 224

Pines, p. 166

Moist Woods

The majority of our area is either forested or would revert to forest
if left undisturbed. Spruce–fir (boreal) woods occur along the
northern edge of our range and the crest of the Appalachians. In a

transition zone between this and the broad-leaved forests are mixed deciduous/evergreen woods, characterized by Eastern Hemlock (p. 164), American Beech (p. 202), Sugar Maple (p. 176), Yellow Birch (p. 200), and White Pine (p. 166). In the broad-leaved deciduous woods, different forest types include beech-maple, oak-hickory, and sugar maple-basswood. The southeastern evergreen forests are typically composed of pines.

The species listed below are found in the broad-leaved deciduous and mixed deciduous/evergreen woods that dominate most of our region. For edible plants found in Spruce-Fir Woods see p. 281; species found in the dry portions of coastal southeastern evergreen forests are listed under Dry Open Woods (see p. 272).

Climate, topography (drainage conditions), and substrata determine the character of any given forest. For example, forests growing over limestone substrata have lusher flora than those growing over granite. Likewise, woods that form in valleys, such as beech-maple, have moister, more fertile soils than woods found higher up on the slopes, such as oak-hickory. The vegetation found on the forest floor is much richer in heavily shaded forests. Species such as Spring-beauties (pp. 32, 104) and Toothworts (pp. 28, 100) are typical of beech-maple woods; Wintergreen (pp. 36, 224) and Greenbriers (pp. 148, 196) commonly occur in oak-hickory woods.

Unlike dry open woods, moist woods have few breaks in the canopy (usually only where trees have fallen), and the forest floor is therefore shady and moist. Where there are sunlit openings, however, variations in plant life do occur. Thus in oak-hickory woods, species more typical of dry open woods may be found. This is also true in isolated areas where the soil is poorer and the topography promotes rapid runoff. Within a given forest the locations of specific plants will vary considerably. For example, the south-facing slope of a ravine receives more sunlight and is likely to support wildflowers that bloom in the spring. On the other hand, the north-facing slope is most likely to produce wildflowers later on in the summer.

Spring

False Solomon's-seal, p. 52
Bellworts, p. 76
Solomon's-seals, pp. 76, 148
Greenbriers, pp. 148, 196

Carrion-flower, p. 148
**Ostrich Fern (north),
  p. 232**

Wild Ginger, pp. 96, 160
Spiderwort, p. 130
Violets, p. 132

Pines, p. 166
Sweetgum, p. 214

Wood-sorrels, pp. 30, 72,
  104, 134

Greenbriers, pp. 148, 196

Trilliums, pp. 24, 96, 146
Star Chickweed, p. 34
Honewort, p. 40
Waterleafs, pp. 44, 128
Cleavers, p. 50
Clintonia (north), pp. 74, 148
Trout-lily, p. 74
Jewelweeds, pp. 78, 92
Wild Lettuces, p. 86

Honewort, p. 40
Dwarf Ginseng (north), p. 42
Harbinger-of-spring, p. 42
Wild Leek, p. 52
Trout-lily, p. 74
Puttyroot, p. 78

Carrion-flower, p. 148
Greenbriers, pp. 148, 196
Eastern Hemlock (north), p. 164

Carrion-flower, p. 148

Wild Leek, p. 52
Indian Cucumber-root, p. 74

Spring-beauties, pp. 32, 104

Trilliums, pp. 24, 96, 146
Spring Cress, p. 28
Toothworts, pp. 28, 100
Wood-sorrels, pp. 30, 72, 104, 134
Star Chickweed, p. 34
Dwarf Ginseng (north), p. 42
Harbinger-of-spring, p. 42
Cleavers, p. 50
False Solomon's-seal, p. 52
Wild Leek, p. 52
Clintonia (north), pp. 74, 148
Indian Cucumber-root, p. 74

Dayflowers, p. 130
Spiderwort, p. 130
Violets, p. 132
Large-leaved Aster (north), p. 144
Blue Lettuces, p. 144
Greenbriers, pp. 148, 196
Wood-nettle, p. 150
Skunk Cabbage (north), p. 156

Wild Onions, p. 114
Redbud, pp. 122, 192
Wild Hyacinth, p. 136
Pines, p. 166
Mulberries, p. 210
Rock Tripe (north), p. 236
Common Morel, p. 238

Pines, p. 166
(Sweet) Birches (north), p. 200
Slippery Elm, p. 200

Greenbriers, pp. 148, 196

Wild Onions, p. 114
Redbud, pp. 122, 192

Solomon's-seals, pp. 76, 148

Solomon's-seals, pp. 76, 148
Wild Lettuces, p. 86
Wild Onions, p. 114
Redbud, pp. 122, 192
Dayflowers, p. 130
Spiderwort, p. 130
Violets, p. 132
Blue Lettuces, p. 144
Twisted-stalk (north), p: 148
Greenbriers, pp. 148, 196
Basswood, p. 194
Ostrich Fern (north), p. 232

Spring Cress, p. 28
Toothworts, pp. 28, 100
Honewort, p. 40
Sweet Cicely, p. 42
Wild Leek, p. 52

Wild Ginger, pp. 96, 160
Wild Onions, p. 114
Redbay (coastal south),
p. 208

Maples, p. 176
Walnuts, p. 188
Hickories, p. 190

(Sweet) Birches (north),
p. 200
Sycamore, p. 214

Wintergreen (north), pp.
36, 224
Spikenard, p. 40
Violets, p. 132
Wood-nettle, p. 150
Eastern Hemlock (north),
p. 164

Pines, p. 166
Yaupon (coastal south),
p. 192
(Sweet) Birches (north),
p. 200
Slippery Elm, p. 200
Common Spicebush, p. 208

Summer

Greenbriers, pp. 148, 196

Wild Ginger, pp. 96, 160
Spiderwort, p. 130
Honey Locust, p. 184

Hazelnuts, p. 200
Sweetgum, p. 214

Cleavers, p. 50

May-apple, p. 20
Wood-sorrels, pp. 30, 72,
104, 134

Greenbriers, pp. 148, 196
Red Mulberry, p. 210

Star Chickweed, p. 34
Dayflowers, p. 130

Greenbriers, pp. 148, 196
Wood-nettle, p. 150

Wild Leek, p. 52
Wild Lettuces, p. 86
Wild Onions, p. 114
Redbud, pp. 122, 192

Blue Lettuces, p. 144
Rock Tripe (north), p. 236
Sulphur Shelf, p. 238

Carrion-flower, p. 148
Greenbriers, pp. 148, 196

Hazelnuts, p. 200

Juneberries, pp. 18, 220
Bunchberry (north), p. 20
May-apple, p. 20
Wood Strawberry
(north), p. 30

Wild-raisins, p. 178
Mulberries, p. 210
Gooseberries and Cur-
rants, p. 212
Blueberries, p. 220

Juneberries, pp. 18, 220
May-apple, p. 20
Wood Strawberry
(north), p. 30
Carrion-flower, p. 148
Greenbriers, pp. 148, 196

Wild-raisins, p. 178
Red Mulberry, p. 210
Gooseberries and Cur-
rants, p. 212
Blueberries, p. 220

Hazelnuts, p. 200

Wild Leek, p. 52
Indian Cucumber-root,
p. 74

Wild Onions, p. 114

Solomon's-seals, pp. 76, 148

Wood-sorrels, pp. 30, 72,
104, 134
Star Chickweed, p. 34
Wild Leek, p. 52
Indian Cucumber-root,
p. 74

Wild Onions, p. 114
Dayflowers, p. 130
Greenbriers, pp. 148, 196

Sweet Cicely, p. 42
Wild Leek, p. 52
Wild Ginger, pp. 96, 160
Wild Onions, p. 114
Common Wax-myrtle
(south), p. 206

Redbay (coastal south),
p. 208
Common Spicebush,
p. 208

Wood Strawberry (north),
p. 30
Wintergreen (north), pp.
36, 224
Spikenard, p. 40
Whorled Loosestrife, p. 72
Bee-balm, p. 118
Wood-nettle, p. 150

Pines, p. 166
Yaupon (coastal south),
p. 192
Basswood, p. 194
(Sweet) Birches (north),
p. 200
Common Spicebush, p. 208

Fall

Wild Ginger, pp. 96, 160
Walnuts, p. 188
Hickories, p. 190
Hazelnuts, p. 200

Chestnut, p. 202
Oaks, p. 204
Sweetgum, p. 214

American Beech, p. 202

Yellow Wood-sorrels, p. 72

Greenbriers, pp. 148, 196

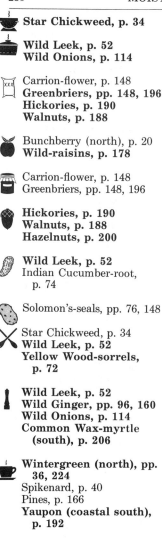

Star Chickweed, p. 34

Wild Leek, p. 52
Wild Onions, p. 114

Rock Tripe (north), p. 236
Sulphur Shelf, p. 238

Carrion-flower, p. 148
Greenbriers, pp. 148, 196
Hickories, p. 190
Walnuts, p. 188

Hazelnuts, p. 200
American Beech, p. 202
Chestnut, p. 202
Oaks, p. 204

Bunchberry (north), p. 20
Wild-raisins, p. 178

Pawpaw, p. 192
American Hackberry, p. 194

Carrion-flower, p. 148
Greenbriers, pp. 148, 196

Wild-raisins, p. 178

Hickories, p. 190
Walnuts, p. 188
Hazelnuts, p. 200

American Beech, p. 202
Chestnut, p. 202
Oaks, p. 204

Wild Leek, p. 52
Indian Cucumber-root,
   p. 74

Wild Onions, p. 114

Solomon's-seals, pp. 76, 148

Star Chickweed, p. 34
Wild Leek, p. 52
Yellow Wood-sorrels,
   p. 72

Indian Cucumber-root,
   p. 74
Wild Onions, p. 114

Wild Leek, p. 52
Wild Ginger, pp. 96, 160
Wild Onions, p. 114
Common Wax-myrtle
   (south), p. 206

Redbay (coastal south),
   p. 208
Common Spicebush,
   p. 208

Wintergreen (north), pp.
   36, 224
Spikenard, p. 40
Pines, p. 166
Yaupon (coastal south),
   p. 192

(Sweet) Birches (north),
   p. 200
Common Spicebush, p. 208

Fall–Early Spring

Sweetgum, p. 214

Kentucky Coffee-tree,
p. 180

 Greenbriers, pp. 148, 196

 Wild Leek, p. 52   **Hog-peanut, p. 124**
Wild Onions, p. 114   Rock Tripe (north), p. 236

 Skunk Cabbage (north),   Carrion-flower, p. 148
p. 156   **Greenbriers, pp. 148, 196**
Jack-in-the-pulpit, p. 156

**Wintergreen (north), pp.
36, 224**

Carrion-flower, p. 148   Greenbriers, pp. 148, 196

Wild Leek, p. 52   **Wild Onions, p. 114**

Solomon's-seals, pp. 76, 148

Wintergreen (north), pp. 36,   **Wild Leek, p. 52**
224   **Wild Onions, p. 114**
Partridgeberry, pp. 36, 102,
174

Wild Leek, p. 52   **Redbay (coastal south),**
Wild Onions, p. 114   **p. 208**

Wintergreen (north), pp.   (Sweet) Birches (north),
36, 224**   **p. 200**
Spikenard, p. 40   Common Spicebush, p. 208
Pines, p. 166
**Yaupon (coastal south),
p. 192**

Spruce–Fir (Boreal) Woods

Spruce–fir woods enter our area only around the upper Great
Lakes, in northern New England, and along the high ridge of the
Appalachians. These are thick evergreen forests dominated by
Red Spruce, p. 168, and Balsam Fir, p. 164. (Fraser Fir, *Abies
fraseri,* replaces Balsam Fir in the southern Appalachians.) The
dimly lit forest floor is covered with a thick carpet of needles and
remains cool and somewhat moist even at the hottest times of the
year. Few herbs and shrubs are found here, except in those places
where breaks in the forest canopy allow the passage of sunlight.

Spring

Pines, p. 166   **Spruces, p. 168**

Common Wood-sorrel, pp.
30, 104

Clintonia, pp. 74, 148   Large-leaved Aster, p. 144

Pines, p. 166   Spruces, p. 168
Rock Tripe, p. 236

Balsam Fir, p. 164   Yellow Birch, p. 200
Pines, p. 166   Reindeer Moss, p. 236
Spruces, p. 168

Common Wood-sorrel,   Clintonia, pp. 74, 148
  pp. 30, 104   Twisted-stalk, p. 148
Trailing Arbutus, pp. 36,
102, 224

 Yellow Birch, p. 200

Labrador-tea, pp. 18, 208   Pines, p. 166
Creeping Snowberry, pp.   **Yellow Birch, p. 200**
  36, 222

Summer

Spruces, p. 168

Common Wood-sorrel, pp.
30, 104

Roseroot, p. 112

Rock Tripe, p. 236

Reindeer Moss, p. 236

Juneberries, pp. 18, 220   **Wild-raisins, p. 178**
Bunchberry, p. 20   **Gooseberries and Cur-**
Creeping Snowberry, pp. 36,   **rants, p. 212**
222   **Blueberries, p. 220**
Bearberry, pp. 36, 224

Juneberries, pp. 18, 220   Wild-raisins, p. 178
Gooseberries and Cur-   **Blueberries, p. 220**
  rants, p. 212

**Common Wood-sorrel,
pp. 30, 104**
Roseroot, p. 112

Twisted-stalk, p. 148

Labrador-tea, pp. 18, 208
**Creeping Snowberry, pp.
36, 222**

Pines, p. 166
Yellow Birch, p. 200

Fall

Spruces, p. 168

Rock Tripe, p. 236

Reindeer Moss, p. 236

Bunchberry, p. 20
Bearberry, pp. 36, 224

**Wild-raisins p. 178
Mountain-ashes, p. 188**

 Wild-raisins, p. 178

Mountain-ashes, p. 188

Labrador-tea, pp. 18, 208
**Creeping Snowberry, pp.
36, 222**

Pines, p. 166
Yellow Birch, p. 200

Fall–Early Spring

Rock Tripe, p. 236

Reindeer Moss, p. 236

Mountain-ashes, p. 188

Mountain-ashes, p. 188

Partridgeberry, pp. 36, 102,
174

Labrador-tea, pp. 18, 208
**Creeping Snowberry, pp.
36, 222**

Pines, p. 166
Yellow Birch, p. 200

Alpine Tundra

Small areas of alpine tundra can be found at the top of a few of the
higher mountains of northern New England and the Gaspé Peninsula. Although many of the plant species found here are different
from those found in the true tundra far to the north, they endure
virtually the same extremes of cold and exposure.

Vegetation is limited to low grasses and sedges, and ground-hugging shrubs such as Mountain Cranberry (pp. 102, 222) and Black Crowberry (pp. 102, 168.) **Note:** Alpine tundra is a fragile habitat and special care should be taken when entering it. Once the vegetation has been destroyed, it takes a great deal of time to grow back.

Spring

 Rock Tripe, p. 236

Reindeer Moss, p. 236   Iceland Moss, p. 236

Labrador-tea, pp. 18, 208
Creeping Snowberry, pp. 36, 222

Summer

Roseroot, p. 112

Alpine Smartweed, p. 116   Rock Tripe, p. 236

Reindeer Moss, p. 236   Iceland Moss, p. 236

Cloudberry, p. 32   Black Crowberry, pp. 102, 168
Creeping Snowberry, pp. 36, 222   **Blueberries, p. 220**

 Cloudberry, p. 32   **Blueberries, p. 220**

Roseroot, p. 112   Alpine Smartweed, p. 116

Labrador-tea, pp. 18, 208
Creeping Snowberry, pp. 36, 222

Fall

Mountain Cranberry, pp. 102, 222

Alpine Smartweed, p. 116   Rock Tripe, p. 236

Reindeer Moss, p. 236   Iceland Moss, p. 236

Mountain Cranberry, pp. 102, 222   Black Crowberry, pp. 102, 168

 Mountain Cranberry, pp. 102, 222

Alpine Smartweed, p. 116

Labrador-tea, pp. 18, 208
**Creeping Snowberry, pp.
   36, 222**

Fall-Early Spring

**Mountain Cranberry, pp.
   102, 222**

 Rock Tripe, p. 236

Reindeer Moss, p. 236   Iceland Moss, p. 236

**Mountain Cranberry, pp.   Black Crowberry, pp. 102,
   102, 222**   168

**Mountain Cranberry, pp.
   102, 222**

Labrador-tea, pp. 18, 208
**Creeping Snowberry, pp.
   36, 222**

Food Uses

The plants described in this *Field Guide* have been categorized according to similarities in their preparation and use. Twenty-two food-use categories have been created (such as, Pickles, Potatoes, and Salads), generally in terms of familiar domesticated foods, and are discussed in alphabetical order on the following pages. With the exceptions of Vinegars, and Wines and Beers, each category has its own descriptive symbol; these are defined in *Explanation of Symbols* (p. 15). Within each category, species are grouped according to the season of the year in which they become available, with abundant, frequently used species entered in boldface. It should be noted that species included in Fall–Early Spring are available in both the fall and early spring, although they may not be formally listed under those seasons. General geographic designations are indicated through the use of the terms Far North, North, South, and Deep South in parentheses after the species name; see *Finding Edible Plants* (p. 241) for their definition.

Directions for preparation and storage are intended to maximize the retention of nutritive value and minimize the loss of vitamins and minerals. Readers should familiarize themselves with the general instructions given in the applicable food-use description and then refer to the individual species accounts in the main text for details.

In this guide, only the most basic cooking directions are given — sufficient to make the species edible; plants should first be tried using the instructions given here, and then, when taste and texture are familiar, prepared in more elaborate fashion using one of the cookbooks listed in *Recommended Books* (p. 313). Directions are not given for freezing and canning — for these, refer also to the appropriate books in *Recommended Books*. **Warning:** Before preparing any plant see the main text for specific directions. Some plants are edible only at specific times of the year and others are edible only under precise conditions.

Asparagus

The species listed below produce stout shoots which can be cooked and served like asparagus. Gather only the tender young tips, and rub off any hairs or scales. Wash briefly but thoroughly in *cold* water, pat dry, and store in a refrigerator until ready to be cooked. To limit vitamin loss, use as little cooking water as possible, and add salt only when the vegetable is ready to be served. Steaming is preferable to boiling, except in those cases calling for changes of water (for example, Pokeweed, p. 46, Common Milkweed, p. 112, and Sicklepod, p. 82). Avoid overcooking. **Caution:** Shoots are normally gathered at a time when no flowers are present for proper species identification. If you are the least bit uncertain as to the identity of a shoot, do not use it.

Spring

Japanese Knotweed,
  p. 46
Pokeweed, p. 46
False Solomon's-seal, p. 52
Asparagus, p. 76
Bellworts, p. 76
Solomon's-seals, pp. 76, 148
Day-lily, p. 92
Fireweed (north), p. 98
River-beauty (far north),
  p. 98

Common Milkweed,
  p. 112
Carrion-flower, p. 148
Greenbriers, pp. 148, 196
Cattails, pp. 158, 230
Large Cane (south), p. 228
Bracken, p. 232
Ostrich Fern (north),
  p. 232

Summer

Sicklepod, p. 82
River-beauty (far north),
  p. 98

Lesser Broomrape, p. 128
Greenbriers, pp. 148, 196

♉ Candies

Although confections made from wild plants were once an important part of American life, most have ceased to be popular. Only a few, such as candied Violets or Angelica, remain in general usage to remind us of other times. Roots, shoots, and nuts were simmered in a rich sugar syrup until thoroughly saturated, then partially dried, and rolled in granulated sugar (see Wild Ginger, pp. 96 and 160). Flower petals were candied using the method described for Spiderwort, p. 130. Marsh Mallow, p. 108, was the original source for marshmallows.

Spring

Angelica (north), p. 40
Sweetflag, pp. 62, 156
Coltsfoot (north), p. 84
Wild Ginger, pp. 96, 160

Spiderwort, p. 130
Violets, p. 132
Pines, p. 166

Summer

Angelica (north), p. 40
Coltsfoot (north), p. 84
Wild Ginger, pp. 96, 160
Roses, pp. 106, 184
Great Burdock, p. 126

Spiderwort, p. 130
Honey Locust, p. 184
Hazelnuts, p. 200
Reed, p. 228

Fall

Wild Ginger, pp. 96, 160   Chestnut, p. 202
Walnuts, p. 188   **Eastern Chinquapin,**
Hickories, p. 190   **p. 202**
Hazelnuts, p. 200   **Oaks, p. 204**

Fall–Early Spring

Elecampane, p. 88
Marsh Mallow (north),
    p. 108

⚘ Cereals

Only a few plants produce edible seeds in large enough quantities to warrant their collection and use as cereals. Wild Rice, p. 228, is by far the most prized, and, indeed, when properly prepared, it makes one of the finest of all vegetables, wild or domestic. The other wild cereals are not as tasty, but they are just as nutritious. Lamb's-quarters, p. 152, and Reed, p. 228, produce tiny seeds that can be boiled, hull and all, to make a kind of breakfast gruel. Common Sunflower, p. 88, and Large Cane, p. 228, can also be made into gruel, but their seeds must be hulled first. Pickerelweed, p. 136, yields small green fruit which are usually added raw to granolas.

Summer

Common Sunflower (west),   Large Cane (south), p. 228
    p. 88   **Wild Rice, p. 228**
Pickerelweed, p. 136

Fall

Common Sunflower (west),   Lamb's-quarters, p. 152
    p. 88   Reed, p. 228
Pickerelweed, p. 136

⁊ Chewing Gums

A crude chewing gum substitute can be obtained by collecting the hardened sap that forms around wounds in any of the following three species. These native products may be a bit brittle at first.

Summer

Compass-plant (west), p. 88

All Year

Spruces (north), p. 168   **Sweetgum, p. 214**

☕ Coffees

A small group of wild plants produce roots and seeds that can be used as caffeine-free coffee substitutes. Roast in a slow oven — 175°–250°F (79°–121°C) — until dark brown on the inside and then grind. Prepare just as you would store-bought coffee. Make sure your pot is thoroughly clean and always start the brewing process with cold water. Do not boil unless specifically directed to do so. Boiling may not only make the brew bitter, but will drive off some of the volatile oils that contribute to the flavor. Brewing times will, of course, vary according to the species used and your own particular taste. However, they should be kept to a minimum as overbrewing will also make the coffee bitter. As a general rule, use $\frac{1}{2}$ cup of plant material for every 4 cups of water. **Note:** Coffees lose flavor when exposed to air. Keep them in tightly covered containers, especially after they have been ground.

Spring

Water Avens (north),   Chufa, p. 230
p. 94

Summer

Cleavers, p. 50   **Water Avens (north),**
Scotch Broom, pp. 82, 182   **p. 94**
Common Sunflower (west),   Chufa, p. 230
p. 88

Fall

Common Sunflower (west),   Feverwort, p. 160
p. 88   Beech, p. 202
Water Avens (north),   Chufa, p. 230
p. 94

Fall–Early Spring

Common Dandelion,   **Chicory, pp. 58, 144**
p. 84   Oyster-plant (north), p. 126
Yellow Goat's-beard   Kentucky Coffee-tree,
(north), p. 84   p. 180
Water Avens (north),   Chufa, p. 230
p. 94

▤ Cold Drinks

A surprisingly varied assortment of refreshing beverages can be made from wild plants. They range from the very familiar to the entirely new; from the juices squeezed from Cranberries (pp. 102, 222) and Wild Grapes (p. 198), to the lemonadelike drink made by soaking the fruit of Sumacs (p. 186) in cold water, to the unique milky brew extracted from the tubers of Chufa (p. 230), and finally the watery sap of some trees (see Syrups and Sugars, p. 309, for list of species).

Fruit juices in particular are an important source of vitamin C. While a few fruits can be pressed or crushed raw to obtain juice, most should be heated. This improves the extraction process and inactivates the enzymes that would cause the flavor to deteriorate. Adding just enough water to cover, crush the fruit and simmer until soft (do not overcook or the flavor will suffer). Strain the juice through several thicknesses of cheesecloth until completely clear. Dilute and sweeten to taste; drink as is, or for variety, mix with other juices.

If the juice is not going to be used within a few hours, it should be pasteurized. Heat in a double boiler, stirring constantly. As soon as the juice reaches 190°F (88°C) — use a thermometer; below 185°F (85°C) pasteurization is ineffective — put it immediately in the refrigerator, still in the top part of the double boiler. When cold, transfer to bottles and store in the refrigerator (use within a week or so), or put into plastic containers and freeze (will keep for a year). **Note:** Use stainless steel, enamel, or glassware cooking utensils. Iron and copper act as catalysts to destroy vitamin C; fruit acids pit aluminum and dissolve the zinc from galvanized steel.

Spring

Wood-sorrels, pp. 30, 72, 104, 134

Sorrels, pp. 116, 154

Greenbriers, pp. 148, 196

Chufa, p. 230

Summer

Common Elderberry, pp. 18, 172

Mayapple, p. 20

Wood-sorrels, pp. 30, 72, 104, 134

Brambles, pp. 30, 184

Passion-flower (south), p. 94

Sorrels, pp. 116, 154

Greenbriers, pp. 148, 196

Canada Buffaloberry (far north), p. 174

Highbush-cranberry (north) p. 178

Sumacs, p. 186

Red Mulberry, p. 210

Chufa, p. 230

Fall

Yellow Wood-sorrels, p. 72
Passion-flower (south),
  p. 94
Cranberries (north), pp.
  102, 222
Greenbriers, pp. 148, 196
Highbush-cranberry
  (north), p. 178

Common Barberry
  (north), p. 192
Wild Grapes, p. 198
Domestic Apple, p. 216
Chufa, p. 230

Fall–Early Spring

Cranberries (north), pp.
  102, 222

Greenbriers, pp. 148, 196
Chufa, p. 230

☕ Cooked Greens

Although leafy greens will necessarily lose some of their vitamin potency when cooked, this can be kept to a minimum through proper handling and preparation. Treat them like salad greens prior to cooking. Gather only the tender young leaves. (Ideally, they should be used as soon as they are picked.) Wash briefly in cold water to remove any grit, and cook in as little liquid as possible. Often the only water needed will be the moisture retained on the leaves after washing. Steaming is preferable to boiling except for species like Marsh-marigold (p. 70) which require several changes of water in order to remove toxins and render them edible. Do not add salt until the greens are ready to be served.

Keep your cooking times as short as possible. Do not overcook. All vegetables contain aromatic oils that give them their characteristic flavor; these are lost when vegetables are cooked too long. Some cooked greens, such as the Chickweeds (p. 34) and Purslane (p. 72), are bland tasting and should be mixed with strong-flavored greens such as Chicory (pp. 58 and 144) or one of the Mustards (p. 64). Coarse, dry-tasting greens can be improved if salt pork or bacon drippings are added during cooking.

Spring

Water-lilies, p. 22
Trilliums, pp. 24, 96, 146
Field Pennycress, p. 26
Shepherd's Purse, p. 26
Peppergrasses, p. 26
Mallows, pp. 32, 108, 134
Chickweeds, p. 34
Bladder Campion
  (north), p. 34

Caraway (north), p. 38
Honewort, p. 40
Waterleafs, pp. 44, 128
Pokeweed, p. 46
Plantains, p. 46
Saxifrages (north), pp. 48,
  146
Cleavers, p. 50
Sweet Clovers, pp. 56, 80

Summer

Fall

Fall-Early Spring

🍲 Cooked Vegetables

Cooked vegetables such as peas, beans, parsnips, and onions should be prepared in much the same way as cooked greens (see p. 291). Gather them as soon before cooking as possible, and wash briefly but thoroughly in cold water. If the vegetable is not to be cooked immediately, carefully dry it off and store it in the vegetable bin in the refrigerator. Never soak vegetables in water; this dissolves out vitamins, as well as the sugars and aromatic oils that are responsible for the flavor. Do not peel vegetables unless it is absolutely necessary. Peels contribute much to the flavor, and in root vegetables, important minerals are concentrated just under the skin.

As with cooked greens, use as little cooking water as possible, keep cooking times to the barest minimum, and do not add salt until the vegetable is ready to be served. Plan to use any leftover cooking fluid in soups and stews. Obviously, this last suggestion is not applicable to vegetables such as Common Milkweed (p. 112) and Buckbean (p. 22) which require several changes of water to rid the plant of toxins or bitterness. **Note:** A pinch of baking soda will help tenderize fibers, but as this also increases the vitamin loss, it should be reserved for only the most intransigent vegetables.

Spring

Summer

Water-lilies, p. 22
Angelica (north), p. 40
Wild Leek, p. 52
Prickly-pear, p. 60
Yellow Pond-lilies, p. 60
American Lotus, p. 60
Golden Club, p. 62
Wild Mustards, p. 64
Canada Lily, p. 74
Wild Lettuces, p. 86
Day-lily, p. 92
Turk's-cap Lily, p. 92
Sea-rocket, p. 100
**Common Milkweed,
p. 112**
Wild Onions, p. 114
Alpine Smartweed (far
north), p. 116
**Beach Pea (north), pp.
122, 142**

Redbud, pp. 122, 192
Wild Bean, p. 124
Burdocks, p. 126
Thistles, p. 126
Creeping Bellflower (north),
p. 128
Water-hyacinth (deep
south), p. 136
Blue Lettuces, p. 144
Arrow Arum, p. 156
Yuccas, p. 170
**Saw Palmetto (deep
south), p. 170**
Chufa, p. 230
Irish Moss (north), p. 234
Laver, p. 234
Rock Tripe (north), p. 236
Sulphur Shelf, p. 238
Giant Puffball, p. 238

Fall

Wild Leek, p. 52
Yellow Pond-lilies, p. 60
American Lotus, p. 60
Golden Club, p. 62
Day-lily, p. 92
Wild Onions, p. 114
Alpine Smartweed (far
north), p. 116
Wild Bean, p. 124
Thistles, p. 126
Creeping Bellflower (north),
p. 128

Arrow Arum, p. 156
**Saw Palmetto (deep
south), p. 170**
Chufa, p. 230
Irish Moss (north), p. 234
Laver, p. 234
Rock Tripe (north), p. 236
Giant Puffball, p. 238
Shaggy Mane, p. 238
Sulphur Shelf, p. 238

Fall–Early Spring

Buckbean (north), p. 22
Wild Carrot, p. 38
Caraway (north), p. 38
Cow-parsnip (north), p. 40
Water-parsnip, p. 42
Wild Leek, p. 52
Bugleweed, p. 54
**Common Evening-prim-
rose, p. 66**
Wild Parsnip, p. 66

Silverweed (north), p. 70
**Yellow Goat's-beard
(north), p. 84**
Day-lily, p. 92
Marsh Mallow (north),
p. 108
Orpine (north), p. 112
Wild Onions, p. 114
Hyssop Hedge-nettle, p. 120
Woundwort (north), p. 120

▧ Flours

Flour is an excellent way to store food energy for extended periods of time. It can be made from a wide variety of plant parts. Cattail pollen is one of the very best wild flours, richly flavored and readily obtained. The inner bark (cambium) of trees, on the other hand, is difficult to process into flour and not nearly as palatable. However, tree bark can serve as an excellent source of survival food when little else is available.

Drying and washing (leaching) are the two basic methods of preparation. While only a limited number of plants are suited to leaching, flour can be obtained from any plant by drying. Plants should be dried in a warm, well-ventilated place such as an attic, or a *very* slow oven. When thoroughly dry, crush and grind the plants; then sift to remove any fiber. Store the flour in tightly sealed containers. For a complete description of leaching, see Cattails (p. 158).

To process grain into flour, the kernels must first be freed from their papery hulls by winnowing. After the grain has been thoroughly dried, rub it gently between your hands to separate the hulls from the kernels. Then place the material on a sheet and winnow by tossing it repeatedly in the air. The hulls (chaff) are lighter than the kernels, and a gentle wind will blow them away.

An easy way to separate nutmeats from their shells is described in Oils (p. 301). Be sure to crush the nuts thoroughly. Flour so obtained can be used as is or dried for future use. **Note:** Wild flours should be mixed half and half with wheat flour for better cohesion.

Spring

Summer

Clovers, pp. 56, 80, 124
American Lotus, pp. 60
Prickly-pear, p. 60
Yellow Pond-lilies, p. 60
Golden Club, p. 62
Purslane, p. 72
Black Medick, p. 80
**Common Sunflower
(west), p. 88**
Blue Vervain, p. 134
Pickerelweed, p. 136
Carrion-flower, p. 148

Greenbriers, pp. 148, 196
Cattails, pp. 158, 230
Kudzu (south), p. 182
Hazelnuts, p. 200
Large Cane (south), p. 228
Wild Rice, p. 228
Great Bulrush, p. 230
Chufa, p. 230
Reindeer Moss (north),
p. 236
Iceland Moss (far north),
p. 236

Fall

Water-lilies, p. 22
Wild Calla (north), p. 22
Yellow Pond-lilies, p. 60
American Lotus, p. 60
Prickly-pear, p. 60
Golden Club, p. 62
Black Medick, p. 80
**Common Sunflower
(west), p. 88**
Blue Vervain, p. 134
Pickerelweed, p. 136
Carrion-flower, p. 148
Greenbriers, p. 148, 196
Lamb's-quarters, p. 152
Amaranth, p. 154
Kudzu (south), p. 182

Walnuts, p. 188
Hickories, p. 190
Hazelnuts, p. 200
American Beech, p. 202
Chestnut, p. 202
**Eastern Chinquapin,
p. 202**
Oaks, p. 204
Reed, p. 228
Chufa, p. 230
Great Bulrush, p. 230
Reindeer Moss (north),
p. 236
Iceland Moss (far north),
p. 236

Fall–Early Spring

Wild Calla (north), p. 22
Buckbean (north), p. 22
Golden Club, p. 62
Water-shield, p. 96
Skunk Cabbage (north),
p. 156
Jack-in-the-pulpit, p. 156
Arrow Arum, p. 156
Carrion-flower, p. 148
Greenbriers, pp. 148, 196

Cattails, pp. 158, 230
Kudzu (south), p. 182
Reed, p. 228
Great Bulrush, p. 230
Chufa, p. 230
Reindeer Moss (north),
p. 236
Iceland Moss (far north),
p. 236

Fritters

Several plants produce flowers that can be made into delicate-tasting fritters. Essentially, this entails dipping the flowers in a batter made from eggs and flour and then deep-fat frying them in a light oil (peanut oil is best). An excellent recipe for batter can be found in *Joy of Cooking* (see *Recommended Books,* p. 313). Gather blossoms early in the day while still fresh and fragrant, wash briefly in cold water, thoroughly dry, and use them as soon as possible. Common Milkweed flowers (p. 112) differ from the others in that they must be blanched for one minute before being dipped in batter.

Spring

Common Dandelion,   Wisteria (south), p. 180
  p. 84   **Black Locust, p. 184**

Summer

Common Elderberry, pp.   **Day-lily, p. 92**
  18, 172   Common Milkweed, p. 112
Common Dandelion,
  p. 84

Fruit

During the late summer and fall, fruit is one of the most readily obtainable wild foods. It is abundant, easy to find and identify, and requires little preparation. Rich in nutritive value, fruit contains significant amounts of vitamins A, B_2 (riboflavin), and C. Methods of storage and preparation are familiar and will not be discussed here, with the exception of drying techniques.

Fruit can be dried either outdoors on a window screen or indoors on trays in the oven. Collect the fruit when fully ripe, remove seeds and stems, and cut into pieces if necessary. Spread in a single layer on the screen or tray.

Outside, the drying will take a week to 10 days; in the oven, only 6–24 hrs. Fruit dried outdoors should not be placed in direct sunlight; in the oven, set the temperature at 165°F (74°C); with older ovens this may involve the use of a thermometer and some improvisation. In the oven, check the fruit regularly to make sure it is drying evenly and not too quickly (it should feel moist, and cooler than the oven air). Outdoors, beware of rain, wind, dogs, and other mishaps.

The fruit is ready if it leaves no moisture on the skin when squeezed, and springs back to its original shape when released. To equalize the moisture content and improve the texture, "sweat" the dried fruit in tightly sealed containers for a week. Then store in *airtight* containers in a dark place or in the refrigerator.

To reconstitute, add just enough water to cover, and soak until the fruit returns to its original size and appearance. When cooking fruit, use the water in which it was soaked. **Note:** Dried fruit has a higher sugar content than fresh; taste before sweetening. Also, use only stainless steel, enamel, or glass cookware.

Summer

Common Elderberry, pp. 18, 172
Juneberries, pp. 18, 220
Bunchberry (north), p. 20
Mayapple, p. 20
Strawberries, p. 30
Brambles, pp. 30, 184
Cloudberry (far north), p. 32
Creeping Snowberry (north), pp. 36, 222
Bearberry (north), pp. 36, 224
Prickly-pear, p. 60
Ground-cherries, p. 68
Northern Fly-honeysuckle (north), pp. 78, 174
Passion-flower (south), p. 94

Black Crowberry (far north), pp. 102, 168
Roses, pp. 106, 184
Purple-flowering Raspberry (north), pp. 106, 212
Strawberry-blite (north), p. 152
Canada Buffaloberry (far north), p. 174
Wild-raisins, p. 178
Highbush-cranberry (north), p. 178
Mulberries, p. 210
Gooseberries and Currants, p. 212
Wild Cherries, p. 218
Wild Plums, p. 218
Blueberries, p. 220

Fall

Bunchberry (north), p. 20
Bearberry (north), pp. 36, 224
Prickly-pear, p. 60
Ground-cherries, p. 68
Passion-flower (south), p. 94
Black Crowberry (far north), pp. 102, 168
Cranberries (north), pp. 102, 222
Roses, pp. 106, 184
Wild-raisins, p. 178
Highbush-cranberry (north), p. 178

Mountain-ashes (north), p. 188
Common Barberry (north), p. 192
Pawpaw, p. 192
Persimmon, p. 194
American Hackberry, p. 194
Wild Grapes, p. 198
Domestic Apple, p. 216
Domestic Pear, p. 216
Wild Cherries, p. 218
Wild Plums, p. 218
Chokeberries, p. 220

Fall–Early Spring

Wintergreen (north), pp. 36, 224

Mountain-ashes (north), p. 188

🫙 Jams and Jellies

The secret of jelly-making is in the proper combination of fruit, pectin, acid, and sugar. Pectin and acid are contained in all fruit, but to varying degrees; different amounts can be found in the same fruit at different times of the season and in fruit from the same plant from year to year.

The first step in making jelly is obtaining fruit juice. Wash the fruit thoroughly in cold water, but do not soak. Then follow the procedure described in Cold Drinks (p. 290). Most fruit require heating to help in the extraction process.

The easiest way to make jelly is by adding commercial pectin. Cooking times are predictable, and the results are uniformly good. Be sure to follow the directions that come with whatever type of pectin you use (liquid or powder).

Without added pectin, the process is much more involved. Take one tablespoon of fruit juice (cool, not hot) and combine it with 1 tablespoon of grain or denatured alcohol. If a single jellylike mass forms, the juice is extremely rich in pectin, and one cup of sugar should be added for every cup of juice when making jelly. The formation of several smaller lumps indicates that the juice is only moderately rich in pectin; add $\frac{3}{4}$ cup sugar for each cup of juice. If the juice does not congeal at all, then the amount of pectin is inadequate for making jelly. Either commercial pectin must be added or the juice mixed with that of a pectin-rich fruit, such as Crabapples (p. 216), Wild Grapes (p. 198), or Common Barberry (p. 192).

Once pectin and sugar are present in the proper amounts, bring the mixture to a rapid boil. The best method for testing whether it is ready to jell is by using a thermometer. The jelly is ready to set when its temperature reaches 8°F (5°C) above the boiling point of water (this varies according to altitude). A more traditional but less precise method is as follows: Dip a spoon in the mixture and, holding it away from the heat, tip it sideways. If two drops form, then merge and slide off the spoon in a sheet, the jelly is ready.

Jelly must be stored in sterilized, airtight containers. Wash jars and place them upside down in a pot of boiling water for 10 min. to sterilize. Fill with jelly within $\frac{1}{2}$ in. of the top. To seal, pour $\frac{1}{4}$ in. of parafin on top (parafin should be melted in a double boiler). Prick any air bubbles that form in the parafin in order to insure a perfect seal. When the parafin has hardened, cap the jars tightly to prevent damage by mice.

Jams are made in much the same way as jellies, except that the fruit is not strained out. **Note:** Use stainless steel, enamel, or glass cookware. Iron and copper act as catalysts to destroy vitamin C; fruit acids pit aluminum and dissolve zinc from galvanized steel.

Spring

Japanese Knotweed, p. 46
Carrion-flower, p. 148

Greenbriers, pp. 148, 196
Irish Moss (north), p. 234

Summer

Common Elderberry, pp. 18, 172
Juneberries, pp. 18, 220
Mayapple, p. 20
Strawberries, p. 30
Brambles, pp. 30, 184
Cloudberry (far north), p. 32
Ground-cherries, p. 68
Northern fly-honeysuckle (north), pp. 78, 190
Passion-flower (south), p. 94
Roses, pp. 106, 184

Canada Buffaloberry (far north), p. 174
Wild-raisins, p. 178
Highbush-cranberry (north), p. 178
Carrion-flower, p. 148
Greenbriers, pp. 148, 196
Red Mulberry, p. 210
Gooseberries and Currants, p. 212
Wild Cherries, p. 218
Wild Plums, p. 218
Blueberries, p. 220
Irish Moss (north), p. 234

Fall

Ground-cherries, p. 68
Passion-flower (south), p. 94
Cranberries (north), pp. 102, 222
Roses, pp. 106, 184
Carrion-flower, p. 148
Greenbriers, pp. 148, 196
Highbush-cranberry, (north), p. 178
Wild-raisins, p. 178
Mountain-ashes (north), p. 188

Common Barberry (north), p. 192
Persimmon, p. 194
Wild Grapes, p. 198
Crabapples, p. 216
Domestic Apple, p. 216
Hawthorns, p. 216
Wild Cherries, p. 218
Wild Plums, p. 218
Chokeberries, p. 220
Irish Moss (north), p. 234

Fall-Early Spring

Cranberries (north), pp. 102, 222
Greenbriers, pp. 148, 196
Carrion-flower, p. 148

Mountain-ashes (north), p. 188
Persimmon, p. 194
Irish Moss (north), p. 234

🌰 Nuts

The majority of the plants discussed in *this guide* fall into the fruit and vegetable category of foods. Although they contribute essential vitamins and minerals, they do not provide the proteins and fatty acids necessary for a healthy diet. They should only be thought of as a portion of a meal, not nutritionally complete by themselves. Most nuts, on the other hand, are rich sources of protein and fatty acids. Walnuts, for instance, contain as much as 18% protein and 60% fat, with a single pound of shelled nutmeats supplying up to 3000 calories. Nuts can be substituted for the more conventional sources of protein in a meal, as vegetarians have done for centuries, and because the fatty acids are unsaturated, it is often healthier to do so.

Because nutmeats will become rancid if exposed to light, heat, moisture, and air, nuts should be stored in their own shells whenever possible. Harvest the nuts as soon as they ripen; husk them, and spread in a thin layer in a dark, well-ventilated (dry) place such as an attic. Allow to cure for several days, and then store in baskets or mesh bags in a cool, somewhat humid place such as a basement. These should keep for a year or more if held at 40°F (5°C). Nuts that have been cured, but are already shelled, should be stored in tightly sealed containers in a cool, dark, dry place or in a freezer.

In addition to being eaten raw or roasted, as a snack or as a main dish, nuts can be candied (see Candies, p. 287), or made into flour (see Flours, p. 295) or oil (see Oils, below).

Summer

American Lotus, p. 60
**Common Sunflower
(west), p. 88**

Hazelnuts, p. 200

Fall

American Lotus, p. 60
**Common Sunflower
(west), p. 88**
Walnuts, p. 188
Hickories, p. 190
Hazelnuts, p. 200

**Eastern Chinquapin,
p. 202**
American Beech, p. 202
Chestnut, p. 202
Oaks, p. 204

🌰 Oils

Unsaturated vegetable oils can be extracted from a few of the richer nuts. Crack open the nuts, discard any that are spoiled, and then pound until the nutmeats are fairly well loosened from their shells. Put the fragmented nuts in a bowl of cold water and separate the nutmeats from the shells. (Sunflower shells will float and the nutmeats will sink; with walnuts and hickories, the nutmeats will float and the shells will sink.) Taking only the nutmeats, boil gently. Skim off the oil when it rises to the surface of the water. Store tightly sealed in the refrigerator and use within a few weeks; it quickly goes rancid. **Note:** Nut oils are light-textured and break down under high heat. Use them in salads, *not* to fry foods.

Summer

**Common Sunflower
 (west), p. 88**

Fall

**Common Sunflower   Hickories, p. 190
 (west), p. 88   American Beech, p. 202
Walnuts, p. 188**

🌶 Pickles

Pickling is a time-honored way of preserving fruits and vegetables. There are many different methods, and all can be used successfully with the species listed here. For specific recipes refer to *Recommended Books* (p. 313), in particular *Preserving the Fruits of the Earth* (Schuler and Schuler) and *Putting Food By* (Hertzberg, Vaughan, and Greene).

Many recipes call for soaking the vegetables in either ice water or salted water before pickling. Although this insures crispness, it can result in the loss of many water-soluble vitamins and minerals. In *Let's Cook It Right,* Adelle Davis recommends adding a little calcium salt (calcium chloride) to the water instead of table salt (sodium chloride). Not only does this limit vitamin loss, but it actually adds nutritive value (in the form of calcium). After the vegetables have been soaked, use the water in the final pickling solution.

You can pickle with either white or cider vinegar. The flavor imparted by cider vinegar is milder and better. However, white vinegar is less apt to discolor vegetables and thus is preferable with paler varieties such as Wild Leek (p. 52) and Indian Cucumber-root (p. 74). For more unusual results, use homemade vinegar (see Vinegars, p. 311). **Warning:** If there is the slightest hint of spoilage, do not taste. Discard pickles that are mushy, slimy, moldy, or have an unpleasant odor. Color changes are also an indicator of spoilage.

Spring

Pokeweed, p. 46
Wild Leek, p. 52
**Marsh-marigold (north),
  p. 70**
Indian Cucumber-root,
  p. 74

Wild Onions, p. 114
Redbud, pp. 122, 192
Glassworts, p. 146
Cattails, pp. 158, 230

Summer

Wild Leek, p. 52
Wild Mustards, p. 64
Purslane, p. 72
Indian Cucumber-root,
  p. 74

Scotch Broom, pp. 82, 182
Marsh Mallow (north),
  p. 108
Wild Onions, p. 114
Glassworts, p. 146

Fall

Wild Leek, p. 52
Indian Cucumber-root,
  p. 74

Wild Onions, p. 114
Glassworts, p. 146

Fall–Early Spring

Wild Leek, p. 52
Bugleweed, p. 54
**Jerusalem Artichoke,
  p. 88**
Orpine (north), p. 112

Wild Onions, p. 114
**Hyssop Hedge-nettle,
  p. 120**
Woundwort (north), p. 120

Potatoes

A small but important group of plants produce rootstocks, tubers, or corms that are mild-flavored and can be substituted in recipes calling for potatoes. Although a few species can be used at any time of year, most should be gathered during the period between fall and early spring. It is then that the energy for next year's growth is stored in the form of starch.

Within the fall–early spring collecting period, there is some variation in the quality of the food contained in the underground storage organs. In the fall, virtually all the energy is in the form of starch. By early spring, however, some of the starch has begun to convert to sugars in preparation for new growth. The ideal time to collect these wild potato-like foods is thus a matter of personal taste. Some people prefer the starchiness of roots and tubers gathered in the fall; others prefer the slightly sweeter taste of those gathered in the spring.

Because many of the vitamins and minerals are found near the surface of the skin, most potato substitutes should be cooked with their skins intact (wash thoroughly first). There are, of course, certain exceptions: species such as the Yellow Pond-lilies (p. 60) and Groundnut (pp. 122, 160) have tough outer skins that make unpalatable eating. Most wild potatoes cannot be stored for any length of time. Collect them whenever the ground is unfrozen during the cold-weather months, and use as soon as possible. However, a few species (all terrestrial) can be excavated and transferred to the damp sand of a root cellar, where they may keep for some time. **Note:** Boiled or roasted chestnuts, listed under Nuts (p. 301), closely approximate potatoes in taste and texture.

Spring

Spurge-nettle (deep south), p. 32
Spring-beauties, pp. 32, 104

Solomon's-seals, pp. 76, 148
Groundnut, pp. 122, 160
Cattails, pp. 158, 230

Summer

Spurge-nettle (deep south), p. 32

Solomon's-seals, pp. 76, 148
Groundnut, pp. 122, 160

Fall

Spurge-nettle (deep south), p. 32

Solomon's-seals, pp. 76, 148
Groundnut, pp. 122, 160

Fall–Early Spring

Wild Potato-vine, p. 20
Tuberous Water-lily (north), p. 22
Arrowheads, p. 24
Spurge-nettle (deep south), p. 32
American Lotus, p. 60
Yellow Pond-lilies, p. 60

Solomon's-seals, p. 60
Jerusalem Artichoke, p. 88
Water-shield, p. 96
Groundnut, pp. 122, 160
Great Bulrush, p. 230
Great Bur-reed, p. 230

✕ Salads

This is the largest and nutritionally the most important category. Except for a few of the nuts and fruit, it includes all the tender plants that can be eaten without being cooked. The leafy greens are particularly important. More nutrients are concentrated in dark green leaves than in any other food. Not only are they exceptionally rich in vitamin A, but they contain significant amounts of vitamins C, E, K, and the B complex, as well as iron, copper, magnesium, calcium, and other minerals. Few things are as healthful as a tossed green salad.

Salads must be properly prepared in order to be at their nutritional best and most appetizing. Gather only the tenderest young leaves, making sure to discard any that are damaged or tough. Wash the leaves briefly in cold water and pat dry. If they are not to be used immediately, store them in the vegetable bin of the refrigerator.

In mixing a salad use your imagination. Salad plants offer a wide variety of tastes. Try to put strong-flavored greens in the same salad with mild ones. Once your ingredients are assembled, add one to two tablespoons of oil (preferably an unsaturated vegetable oil) and toss until all surfaces are covered. This will seal out oxygen and moisture, which diminish the vitamin content. Add the vinegar and other seasonings just before serving. **Warning:** Avoid gathering salad plants that grow along heavily traveled roadsides or in polluted water. (Some authors have suggested that aquatic plants can be disinfected by washing or soaking them in water to which a purification tablet such as Halazone has been added. However, Abbott Laboratories — the manufacturer of Halazone — has no experimental evidence to show that this method is effective.)

Spring

Trilliums, pp. 24, 96, 146
Shepherd's Purse, p. 26
Field Pennycress, p. 26
Peppergrasses, p. 26
Horseradish, p. 26
Cuckoo-flower (north), pp. 28, 100
Watercress, p. 28
Mountain Watercress, p. 28
Pennsylvania Bittercress, p. 28
Spring Cress, p. 28
Toothworts, pp. 28, 100
Wood-sorrels, pp. 30, 72, 104, 134
Blackberries, pp. 30, 184
Chickweeds, p. 34
Trailing Arbutus, pp. 36, 102, 224
Caraway (north), p. 38
Harbinger-of-spring, p. 42
Dwarf Ginseng (north), p. 42
Plantains, p. 46

Saxifrages (north), pp. 48, 146
Cleavers, p. 50
False Solomon's-seal, p. 52
Wild Leek, p. 52
Sweet Clovers, pp. 56, 80
Clovers, pp. 56, 80, 124
Ox-eye Daisy, p. 58
Chicory, pp. 58, 144
Sweetflag, pp. 62, 156
Wild Mustards, p. 64
Winter Cress, p. 64
Common Evening-primrose, p. 66
Clintonia (north), pp. 74, 148
Indian Cucumber-root, p. 74
Solomon's-seals, pp. 76, 148
Yellow Goat's-beard (north), p. 84
Common Dandelion, p. 84
Sow-thistles, p. 86

Wild Lettuces, p. 86
Day-lily, p. 92
Water-shield, p. 96
Storksbill, p. 104
Orpine (north), p. 112
Wild Onions, p. 114
Lady's-thumb, p. 116
Sorrels, pp. 116, 154
Redbud, pp. 122, 192
Alfalfa, pp. 124, 142
Burdocks, p. 126
Thistles, p. 126
Oyster-plant (north),
    p. 126
American Brooklime
    (north), p. 128
Dayflowers, p. 130
Spiderwort, p. 130
Violets, p. 132
Corn-salad, p. 134

Blue Lettuces, p. 144
Glassworts, p. 146
Twisted-stalk (north),
    p. 148
Greenbriers, pp. 148, 196
Amaranth, p. 154
Curled Dock, p. 154
Cattails, pp. 158, 230
Saw Palmetto (deep
    south), p. 170
American Basswood, p. 194
Great Bulrush, p. 230
Chufa, p. 230
Ostrich Fern (north),
    p. 232
Bracken, p. 232
Laver, p. 234
Dulse (north), p. 234
Sea Lettuce, p. 234
Edible Kelp (north), p. 234

Summer

Yuccas, pp. 20, 170
Shepherd's Purse, p. 26
Field Pennycress, p. 26
Watercress, p. 28
Wood-sorrels, pp. 30, 72,
    104, 134
Cheeses, pp. 32, 108, 134
Chickweeds, p. 34
Seaside Plantain (north),
    p. 46
Bastard-toadflax, p. 48
Wild Leek, p. 52
Clovers, pp. 56, 80, 124
Pilewort, p. 58
Wild Mustards, p. 64
Purslane, p. 72
Indian Cucumber-root,
    p. 74
European Great Burnet
    (north), p. 94
Virginia Meadow-beauty,
    p. 98
Sea-rocket, p. 100
Roseroot (far north), p. 112
Wild Onions, p. 114

Alpine Smartweed (far
    north), p. 116
Sorrels, pp. 116, 154
Alfalfa, pp. 124, 142
Thistles, p. 126
Great Burdock, p. 126
American Brooklime
    (north), p. 128
Creeping Bellflower (north),
    p. 128
Dayflowers, p. 130
Pickerelweed, p. 136
Glassworts, p. 146
Twisted-stalk (north),
    p. 148
Greenbriers, pp. 148, 196
Amaranth, p. 154
Cattails, pp. 158, 230
Saw Palmetto (deep
    south), p. 170
Chufa, p. 230
Laver, p. 234
Dulse (north), p. 234
Sea Lettuce (north), p. 234
Edible Kelp (north), p. 234

Fall

Watercress, p. 28
Chickweeds, p. 34
Wild Leek, p. 52
**Yellow Wood-sorrels,
p. 72**
Indian Cucumber-root,
p. 74
Wild Onions, p. 114
Alpine Smartweed (far
north), p. 116
Thistles, p. 126
Creeping Bellflower (north),
p. 128

Corn-salad, p. 134
Glassworts, p. 146
Amaranth, p. 154
**Saw Palmetto (deep
south), p. 170**
Chufa, p. 230
Dulse (north), p. 234
Edible Kelp (north), p. 234
Laver, p. 234
Sea Lettuce, p. 234

Fall–Early Spring

Watercress, p. 28
Partridgeberry, pp. 36, 102,
174
Wintergreen (north), pp. 36,
224
Wild Leek, p. 52
Bugleweed, p. 54
Winter Cress, p. 64
**Jerusalem Artichoke,
p. 88**
Day-lily, p. 92
Wild Onions, p. 114

Woundwort (north), p. 120
**Hyssop Hedge-nettle,
p. 120**
Corn-salad, p. 134
**Saw Palmetto (deep
south), p. 170**
Chufa, p. 230
Dulse (north), p. 234
Edible Kelp (north), p. 234
Laver, p. 234
Sea Lettuce, p. 234

❚ Seasonings/Condiments

Included here are plants that are used either as spices and flavoring agents, like bay leaves and wintergreen, or condiments, like mustard and horseradish. Chutneys, capers, and the like are more closely allied with pickles (see p. 302).

Some wild seasonings closely resemble familiar kitchen spices. Others are totally unfamiliar. Wild seasonings that approximate familiar spices can be used in the same ways. The more unfamiliar seasonings, however, may require considerable experimentation before their virtues are fully realized.

Seasonings can be dried using the methods described for teas (see p. 309). Leaves should be gathered in the morning after the dew has dried but before the heat of the day; at that time they are most flavorful. The leaves of herbaceous plants reach their full potency just before the flowers appear. Once dried, seasonings should be stored in tightly sealed containers and kept in a cool place out of the direct sunlight. To further insure their freshness, they should be kept in bulk form, then ground fine just prior to use.

When cooking, do not overseason. Properly used, seasonings will provide a subtle blend of tastes. Also, high heat drives off the aromatic oils that are responsible for the spices' flavor. (Thus, the *odor* of cooking indicates that the food itself is losing flavor.) **Note:** When substituting fresh herbs in recipes calling for dry, use larger amounts.

Spring

Horseradish, p. 26
Spring Cress, p. 28
Toothworts, pp. 28, 100
Honewort, p. 40
Sweet Cicely, p. 42
Wild Leek, p. 52
Sweet Clovers, pp. 56, 80
Coltsfoot (north), p. 84
Common Tansy, p. 90

Wild Ginger, pp. 96, 160
Wild Onions, p. 114
Redbay (coastal south), p. 208
Sassafras, p. 210
Dulse (north), p. 234
Sea Lettuce, p. 234
Edible Kelp (north), p. 234

Summer

Field Pennycress, p. 26
Peppergrasses, p. 26
Shepherd's Purse, p. 26
Horseradish, p. 26
Caraway (north), p. 38
Cow-parsnip (north), p. 40
Sweet Cicely, p. 42
Wild Leek, p. 52
Mints, pp. 54, 118, 138
Sweet Clovers, pp. 56, 80
Wild Mustards, p. 64
Coltsfoot (north), p. 84
Common Tansy, p. 90
Day-lily, p. 92
Wild Ginger, pp. 96, 160

Wild Onions, p. 114
Basil (north), p. 120
Ground Juniper (north), p. 168
Sweetgale (north), p. 206
Bayberries, p. 206
Redbay (coastal south), p. 208
Common Spicebush, p. 208
Sassafras, p. 210
Dulse (north), p. 234
Sea Lettuce, p. 234
Edible Kelp (north), p. 234

Fall

Field Pennycress, p. 26
Shepherd's Purse, p. 26
Horseradish, p. 26
Caraway (north), p. 38
Wild Leek, p. 52
Wild Ginger, pp. 96, 160
Wild Onions, p. 114
Ground Juniper (north), p. 168

Bayberries, p. 206
Common Spicebush, p. 208
Redbay (coastal south), p. 208
Dulse (north), p. 234
Sea Lettuce, p. 234
Edible Kelp (north), p. 234

Fall-Early Spring

Horseradish, p. 26   Dulse (north), p. 234
Wild Leek, p. 52   Edible Kelp (north), p. 234
Wild Onions, p. 114   Sea Lettuce, p. 234
Redbay (coastal south),
    p. 208

⌇ Syrups and Sugars

Sugar Maple is the best known, most widely used source of syrups and sugars, but several other trees also produce copious flows of sweetish sap. These should be tapped and the sap boiled down using the process described for Sugar Maple on p. 176. Although the concentration of sugar in the other tree saps is lower than in that of Sugar Maple, they produce comparable syrups; it just takes more sap to get the same amount of syrup. In areas where surface water is scarce or polluted, these same tree saps make an excellent source of drinking and cooking water. **Note:** Readers seeking additional information on tapping Sugar Maples should refer to *The Maple Sugar Book* by Helen and Scott Nearing (New York; Schoeken Books, 1970).

Spring

Maples, p. 176   (Sweet) Birches (north),
Walnuts, p. 188   p. 200
Hickories, p. 190   Sycamore, p. 214

☕ Teas

A wide variety of wild teas is available for those seeking a caffeine-free alternative to oriental tea. (Only Yaupon, p. 192, contains caffeine.) As a group, they represent one of the most common and easiest ways of using edible wild plants.

Most of the teas made from the plants listed below are infusions. Infusions are made by pouring hot water over the tea materials and steeping them for 5–15 min. Use 1 tsp. of *dried* material, or 2 tsp. of *fresh* material, for every cup of water. Normally, the longer the material is steeped, the fuller-bodied the tea. However, some of the leafy teas will release tannin and make the tea bitter if they remain in the water too long. Once the tea has been properly infused, the plant material should be strained out and the tea sweetened to taste with honey or sugar.

Plant materials must be dried and stored properly in order to insure their potency. They should never be dried in direct sunlight; too much heat will drive off the volatile oils that are oftentimes responsible for the teas' flavor. Instead they should be dried in warm, shaded, well-ventilated places such as attics or pantries. A common method of drying leaves on long-stemmed plants is to tie the stems into loose bundles and hang them from the ceiling. A simpler procedure, however, is to strip off the leaves and fold them in newspaper; the newspaper keeps the leaves out of direct sunlight and soaks up any moisture given off in the drying process. This is a particularly convenient method to use when traveling. A final drying technique, frequently used for bulkier tea materials (rose hips and the like), is to spread them in a shallow layer on a horizontal windowscreen. The screen should be raised above the floor on two supports in order to insure maximum air circulation. Once a tea has been thoroughly dried, it will keep for months if sealed in airtight containers and kept out of the direct sunlight.

Spring

Labrador-tea (north), pp. 18, 208
Creeping Snowberry (north), pp. 36, 222
Wintergreen (north), pp. 36, 224
Spikenard, p. 40
Clovers, pp. 56, 80, 124
Comfrey, pp. 68, 110, 134
Coltsfoot (north), p. 84
Alfalfa, pp. 124, 142
Violets, p. 132
Gill-over-the-ground, p. 140

Nettles, p. 150
Wood-nettle, p. 150
Eastern Hemlock (north), p. 164
Pines, p. 166
Yaupon (coastal south), p. 192
(Sweet) Birches (north), p. 200
Slippery Elm, p. 200
Common Spicebush, p. 208
Sassafras, p. 210

Summer

New Jersey Tea, pp. 18, 192
Labrador-tea (north), pp. 18, 208
Strawberries, p. 30
Brambles, pp. 30, 184
Creeping Snowberry (north), pp. 36, 222
Wintergreen (north), pp. 36, 224
Yarrow, p. 38
Spikenard, p. 40
Mountain-mints, p. 54
Mints, pp. 54, 118, 138
Catnip, pp. 54, 140
Clovers, pp. 56, 80, 124

Wild Chamomile (north), p. 58
Shrubby Cinquefoil (north), pp. 70, 180
Common Mullein p. 72
Whorled Loosestrife, p. 72
Horsemint, p. 78
Coltsfoot (north), p. 84
Pineapple-weed (north), p. 90
Sweet Goldenrod, p. 90
Fireweed (north), p. 98
Roses, pp. 106, 184
Bergamots, p. 118
Basil (north), p. 120
Dittany, p. 120

Alfalfa, pp. 124, 142
Downy Wood-mint, p. 138
Gill-over-the-ground, p. 140
American Pennyroyal,
    p. 140
Blue Giant Hyssop (north),
    p. 140
Nettles, p. 150
Wood-nettle, p. 150
Mexican-tea, p. 152
Jerusalem-oak, p. 152

Pines, p. 166
**Yaupon (coastal south),
    p. 192**
Persimmon, p. 194
Basswood, p. 194
**(Sweet) Birches (north),
    p. 200**
Sweetgale (north), p. 206
Sweetfern (north), p. 206
Common Spicebush, p. 208
Sassafras, p. 210

Fall

Labrador-tea (north), pp.
    18, 208
**Creeping Snowberry
    (north), pp. 36, 222**
**Wintergreen (north), pp.
    36, 224**
Spikenard, p. 40
Roses, pp. 106, 184
Mexican-tea, p. 152

Jerusalem-oak, p. 152
Pines, p. 166
**Yaupon (coastal south),
    p. 192**
**(Sweet) Birches (north),
    p. 200**
Common Spicebush, p. 208
Sassafras, p. 210
Hawthorns, p. 216

Fall–Early Spring

Labrador-tea (north), pp.
    18, 208
**Creeping Snowberry
    (north), pp. 36, 222**
**Wintergreen (north), pp.
    36, 224**
Spikenard, p. 40

Pines, p. 166
**Yaupon (coastal south),
    p. 192**
**(Sweet) Birches (north),
    p. 200**
Common Spicebush, p. 208
Sassafras, p. 210

Vinegars

Vinegars have been omitted from the uses listed in the main text for much the same reason as wines (see Wines and Beers, p. 312). However, because of the frequency with which vinegar is used, either in pickling recipes or as a seasoning for cooked greens and vegetables, I will discuss briefly how it is made.

Vinegar forms whenever wine is exposed to the open air. The wine becomes contaminated with a bacteria that feeds upon the alcohol produced by the yeast and converts it into acetic acid. When most of the alcohol has been converted, the wine becomes vinegar. It follows, therefore, that any plant that can be made into wine can also be made into vinegar. There is one catch, however. If you use a wine recipe, do not add as much sugar. The more sugar added, the more alcohol the yeast will produce. When the alcohol content is too high, the bacteria that produces vinegar cannot live.

You may at some time wish to try your hand at making herb vinegar. Essentially, all you need do is add your favorite seasoning or combination of seasonings to a bottle of vinegar and let it steep for four weeks. Then strain the vinegar, transfer it to a new sterilized bottle, and seal tightly. Never use more than three tablespoons of fresh herbs for every quart of vinegar. If you add more, you have no guarantee that the vinegar will have enough strength to act as a preservative, and the batch may spoil. **Note:** A simple way to make your own vinegar is to add wine yeast to an open vat of tree sap, cover with a cloth, and then let it stand for a few weeks. Almost invariably it will end up as vinegar (see Syrups and Sugars, p. 309, for list of species).

Wines and Beers

Although many edible wild plants can be made into wine or beer, I have omitted these uses from the discussions in the visual and descriptive text. This is not because of any strong bias on my part, but simply because wine and beer making is much too complex a subject to be treated adequately in the space available. A hasty treatment would only increase the likelihood of unsatisfactory results. I do not want to discourage you; far from it. Wine and beer making is one of the more satisfying pastimes I have encountered. However, rather than discuss it here, I will refer you to a few books on the subject: *The Art of Making Wine* by Stanley F. Anderson and Raymond Hull (New York; Hawthorn Books, 1970); *Winemaking at Home* by Homer Hardwick (New York; Simon and Schuster, 1974); *The Art of Making Beer* by Stanley F. Anderson and Raymond Hull (New York; Hawthorn Books, 1971); *The Wine Art Recipe Book* edited by Paul Hasler (Winnipeg, Manitoba; Frieser & Son, 1977). **Note:** Before making wine, you must obtain 2 copies of Form 1541 from the Department of the Treasury — Bureau of Alcohol, Tobacco, and Firearms. When filled out and returned, they permit a head of a household to produce 200 gallons of wine a year. There is no registration fee.

Recommended Books

The following list is not meant to be comprehensive. Rather, these are books that I personally found useful when working with edible plants.

Technical Manuals

Fernald, Merritt Lyndon. 1950. *Gray's Manual of Botany*. 8th ed. New York: Van Nostrand Reinhold Co.

Gleason, Henry A. 1952. *The New Britton and Brown Illustrated Flora*. Lancaster, Penn.: Lancaster Press.

Radford, Albert E., Harry E. Ahles, and C. Ritchie Bell. 1968. *Manual of the Vascular Flora of the Carolinas*. Chapel Hill: University of North Carolina Press.

Steyermark, Julian A. 1963. *Flora of Missouri*. Ames, Iowa: Iowa State University Press.

Identification Guides

Arnold, Augusta Foote. 1968. *The Seabeach at Ebb-Tide*. New York: Dover Publications.

Brockman, C. Frank. 1968. *Trees of North America*. New York: Golden Press.

Cobb, Boughton. 1963. *A Field Guide to the Ferns*. Boston: Houghton Mifflin Co.

Grimm, William Carey. 1966. *How to Recognize Shrubs*. Harrisburg, Penn.: Stackpole Books.

Miller, Orson K., Jr. 1972 *Mushrooms of North America*. New York: E. P. Dutton and Co.

Peterson, Roger Tory, and Margaret McKenny. 1968. *A Field Guide to Wildflowers*. Boston: Houghton Mifflin Co.

Petrides, George A. 1972. *A Field Guide to Trees and Shrubs*. 2nd. ed. Boston: Houghton Mifflin Co.

Smith, Alexander H. 1971. *The Mushroom Hunter's Field Guide*. Revised and Enlarged. Ann Arbor: University of Michigan Press.

Zim, Herbert S., and Lester Ingle. 1955. *Seashores*. New York: Simon and Schuster.

Poisonous Plants

Hardin, James W., and Jay M. Arena. 1974. *Human Poisoning from Native and Cultivated Plants*. 2nd. ed. Durham: Duke University Press.

Kingsbury, John M. 1964. *Poisonous Plants of the United States and Canada*. Englewood Cliffs, N.J.: Prentice-Hall.

————. 1965. *Deadly Harvest: A Guide to Common Poisonous Plants*. New York: Holt, Rinehart and Winston.

Edible Wild Plants

Angier, Bradford. 1974. *Field Guide to Edible Wild Plants*. Harrisburg: Stackpole Books.

Berglund, Berndt, and Clare E. Bolsby. 1971. *The Edible Wild*. Toronto: Pagurian Press.

Brackett, Babette, and Maryann Lash. 1975. *The Wild Gourmet*. Boston: David R. Godine.

Fernald, Merritt Lyndon, and Alfred Charles Kinsey, as revised by Reed C. Rollins. 1958. *Edible Wild Plants of Eastern North America*. New York: Harper and Row.

Gibbons, Euell. 1962. *Stalking the Wild Asparagus*. New York: David McKay Co.

————. 1964. *Stalking the Blue-eyed Scallop*. New York: David McKay Co.

————. 1966. *Stalking the Healthful Herbs*. New York: David McKay Co.

Harrington, H. D. 1967. *Edible Native Plants of the Rocky Mountains*. Albuquerque: University of New Mexico Press.

Harris, Ben Charles. 1968. *Eat the Weeds*. Barre, Mass.: Barre Publishers.

Kirk, Donald R. 1970. *Wild Edible Plants of the Western United States*. Healdsburg, Calif.: Naturegraph Publishers.

Krochmal, Connie, and Arnold Krochmal. 1974. *A Naturalist's Guide to Cooking with Wild Plants*. New York: Quadrangle.

Morton, Julia F. 1963. *Wild Plants for Survival in South Florida*. Miami: Hurricane House.

Tatum, Billy Joe. 1976. *Billy Joe Tatum's Wild Foods Cookbook and Field Guide*. New York: Workman Publishing Co.

Cooking and Preservation (General)

Claiborne, Craig. 1961. *The New York Times Cook Book*. New York: Harper and Row.

Davis, Adelle. 1970. *Let's Cook It Right*. New York: The New American Library.

Farmer, Fannie M. 1973. *The Fannie Farmer Cook Book*. New York: Lauter Levin Associates.

Hertzberg, Ruth, Beatrice Vaughan, and Janet Greene. 1974. *Putting Food By*. Brattleboro, Vt.: The Stephen Greene Press.

Rombauer, Irma S., and Marion Rombauer Becker. 1964. *Joy of Cooking*. Toronto: Bobbs-Merrill Co.

Schuler, Stanley, and Elizabeth Meriwether Schuler. 1973. *Preserving the Fruits of the Earth*. New York: The Dial Press.

Index

Notes

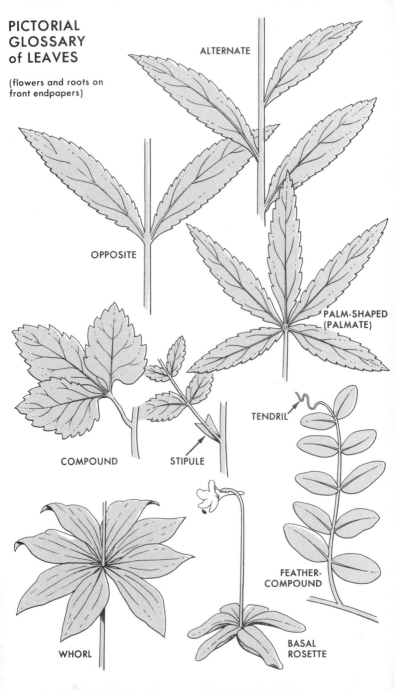

PICTORIAL GLOSSARY of LEAVES

(flowers and roots on front endpapers)

ALTERNATE

OPPOSITE

PALM-SHAPED (PALMATE)

COMPOUND

STIPULE

TENDRIL

WHORL

BASAL ROSETTE

FEATHER-COMPOUND